Thirteen Papers Translated from the Russian

AMERICAN MATHEMATICAL SOCIETY
TRANSLATIONS
Series 2
Volume 144

Thirteen Papers Translated from the Russian

by

S. D. Berman
K. Buzashi [Buzási]
V. K. Dubovoĭ
S. Yu. Favorov
I. S. Kats
I. V. Lobarev
V. È. Lyantse
Yu. I. Lyubarskiĭ
Kh. B. Maĭorga
A. A. Nersesyan
N. P. Pustovoĭtov
G. N. Zholtkevich

AMERICAN MATHEMATICAL SOCIETY
PROVIDENCE, RHODE ISLAND

Translation edited by BEN SILVER

1980 *Mathematics Subject Classification* (1985 *Revision*). Primary 20C07, 20C32, 30D15, 30D20, 30D55, 30E10, 32A15, 34D20, 35J05, 35J25, 45C05, 45K05, 46E15, 46L10, 65F15, 65H10; Secondary 15A52, 46L35, 47F05, 65H15.

Library of Congress Cataloging-in-Publication Data
Thirteen papers translated from the Russian/by S. D. Berman...[et al.; translation edited by Ben Silver].
p. cm. – (American Mathematical Society translations, ISSN 0065-9290; ser. 2, v. 144)
Includes bibliographies.
ISBN 0-8218-3125-9
1. Mathematics. I. Berman, S. D. (Samuil Davidovich), 1922–1987. II. Silver, Ben. III. Title: 13 papers translated from the Russian. IV. Series.
QA3.A572 ser. 2, vol. 144
510 s–dc20 89-15010
[510] CIP

The paper used in this book is acid-free and falls within the guidelines
established to ensure permanence and durability. ♾
This publication was typeset using AMS-TEX,
the American Mathematical Society's TEX macro system.

Contents

Russian Contents*

---

*The American Mathematical Society scheme for transliteration of Cyrillic may be found at the end of index issues of *Mathematical Reviews.*

С. Ю. Фаворов, О целых функциях вполне регулярного роста многих переменных, Теория функций, функц. анализ и их прил. (Харьков), 1982, № 38, 103–111

В. Э. Лянце и Х. Б. Майорга, К теории одноточечной краевой задачи для оператора Лапласа. I, Теория функций, функц. анализ и их прил. (Харьков), 1982, № 38, 84–91

В. Э. Лянце и Х. Б. Майорга, К теории одноточечной краевой задачи для оператора Лапласа. II, Теория функций, функц. анализ и их прил. (Харьков), 1982, № 39, 87–94

Г. Н. Жолткевич, О когомологиях алгебр фон Неймана, Теория функций, функц. анализ и их прил. (Харьков), 1982, № 38, 122–125

Н. А. Пустовойтов, О приближенных алгебраических критериях устойчивости, Числ.-анал. методы исслед. динам. и устойчивости слож. систем. Киев, 1984, 102–110

И. В. Лобарев, Собственные значения и собственные функции уравнения Пайерлса, ВЦ СО АН СССР. Препр., 1984, № 109, 34 с.

Amer. Math. Soc. Transl.
(2) Vol. **144**, 1989

On Modules over Group Algebras of Groups Containing an Infinite Cyclic Subgroup of Finite Index

S. D. BERMAN AND K. BUZÁSI

Dedicated to the memory of László Rédei

§1.

Let G be a group containing an infinite cyclic subgroup of finite index. In this paper we wish to study the representations of G over an arbitrary field K with some restriction on its characteristic. The infinite dihedral group is a particular case of the class of groups considered here. The representations of the infinite dihedral group over an arbitrary field K with $\operatorname{Char} K \neq 2$ are investigated in [1]. The representations of G over an algebraically closed field were considered in [2].

First of all we describe the structure of the group G. It is easy to see that G contains an infinite cyclic normal subgroup (c) of finite index. Let H be the centralizer of the element c in G. Then H has index 2^α $(\alpha = 0, 1)$ in G. Because the subgroup (c) lies in the center of H and has finite index in H, by a known theorem (see [3]) the commutator subgroup of H is finite. It easily follows from this that H can be expressed in a semidirect product $H = H_1 \cdot (a)$, where H_1 is a finite normal subgroup of H and (a) an infinite cyclic group. Since $(G : H) = 2^\alpha$ $(\alpha = 0, 1)$, it is easy to see that G contains a finite normal subgroup F such that G/F is either the infinite cyclic group or the infinite dihedral group:

$$G = F \cdot (a) \quad \text{or} \quad G = (F, a, b), \tag{1}$$

where (a) is an infinite cyclic group and

$$b^{-1}ab = f_1 a^{-1}, \quad b^2 = f_2, \qquad f_1, f_2 \in F. \tag{2}$$

1980 *Mathematics Subject Classification* (1985 *Revision*). Primary 20C07, 20C32.
Translation of Studia Sci. Math. Hungar. **16** (1981), 455–470 (1983); MR **85g**:20011.

We shall investigate finitely generated modules over the group algebra KG, assuming that the characteristic of the field K does not divide the order of the group F. All the KG-modules we consider are assumed to be left and unitary.

DEFINITION 1. Let $\mathscr{D}$ be a skew field containing the field K in its center, and $(\mathscr{D} : K) < \infty$. An associative algebra A is called an *algebra of type E over K* if it is a crossed product of the skew field $\mathscr{D}$ with either the infinite cyclic group

$$A = (\mathscr{D}, a), \qquad a^{-1}\lambda a = \lambda^{\varphi}, \tag{3}$$

where φ is a K-automorphism of $\mathscr{D}$, $\lambda \in \mathscr{D}$, and (a) is the infinite cyclic group or the infinite dihedral group

$$A = (\mathscr{D}, a, b), \qquad a^{-1}\lambda a = \lambda^{\varphi},\ b^{-1}\lambda b = \lambda^{\psi},\ b^{-1}ab = \gamma a^{-1},\ b^2 = \mu, \tag{4}$$

where $\lambda, \gamma, \mu \in \mathscr{D}$, and φ and ψ are K-automorphisms of $\mathscr{D}$.

THEOREM 1. *Let G be a group containing an infinite cyclic subgroup of finite index, and K a field whose characteristic does not divide the order of F* (*see* (1) *and* (2)). *Then the group algebra KG is isomorphic to the orthogonal sum of total matrix rings over some algebra of type E over K.*

PROOF. Because of the restriction on the characteristic of K the group algebra KF is an orthogonal sum of simple algebras $KF = V_1 \dotplus \cdots \dotplus V_s$, where the V_i are total matrix rings over some skew fields $\mathscr{D}_i$. Here K lies in the center of each $\mathscr{D}_i$ and $(\mathscr{D}_i : K) < \infty$ $(i = 1, \dots, s)$.

Let e_i be a minimal central idempotent of KF generating the ideal V_i $(i = 1, \dots, s)$. Since F is a normal subgroup of G, the set of idempotents $\{e_i\}$ decomposes into the orbits

$$\{e_{11}, \dots, e_{1r_1}\}, \dots, \{e_{t1}, \dots, e_{tr_t}\},$$

under the action of automorphisms of KF generated by the inner automorphisms of the group. We set $u_i = e_{i1} + \cdots + e_{ir_i}$ $(i = 1, \dots, t)$. Then $u_1, \dots, u_t$ are idempotents of the center of KG and the orthogonal decomposition

$$KG = U_1 \dotplus \cdots \dotplus U_t \tag{5}$$

holds, where

$$U_i = KGu_i = KGe_{i1} + \cdots + KGe_{ir_i} \qquad (i = 1, \dots, t). \tag{6}$$

We shall show that every algebra U_i is isomorphic to a total matrix ring over some algebra A_i of type E over K $(i = 1, \dots, t)$. We denote by H_i the stabilizer of the idempotent e_{i1} in G $(i = 1, \dots, t)$. Then $H_i \supset F$. If G/F is the infinite cyclic group, then H_i/F is also an infinite cyclic group, and if G/F is the infinite dihedral group, then H_i/F is either an infinite cyclic group or the infinite dihedral group.

We show that the algebra KH_ie_{i1} is K-isomorphic to a total matrix ring over some algebra of type E over K. First we consider the most difficult case, when H_i/F is an infinite dihedral group,

$$H_i = (F, a, b), \qquad b^{-1}ab = f_1a^{-1},\ b^2 = f_2, \tag{7}$$

where $f_1, f_2 \in F$ and (a) is the infinite cyclic group. (The elements a, b, f_1, and f_2 in (7) are not necessarily the same as in (2).)

Let $U = KFe_{i1}$. As mentioned above, U is a total matrix ring of order n over some skew field $\mathscr{D}$. Denote by Z the center of $\mathscr{D}$. Every element $x \in U$ can be uniquely expressed in the form $x = \sum \lambda_{ij}u_{ij}$, where $\lambda_{ij} \in \mathscr{D}$ and $\{u_{ij}\}$ is the system of matrix units

$$u_{kl} \cdot u_{rs} = \begin{cases} 0, & l \neq r \\ u_{ks}, & l = r. \end{cases}$$

Here, as we know, the set $\{u_{ij}\}$ uniquely determines the skew field $\mathscr{D}$, because $\mathscr{D}$ consists of all the elements of the algebra U commuting with the elements a and b in (7).

We define two automorphisms $\varphi: x \to a^{-1}xa$ and $\psi: x \to b^{-1}xb$ $(x \in U)$ of the algebra U, where the elements a and b are defined in (7). Let

$$\varphi(u_{ij}) = u'_{ij}, \qquad \psi(u_{ij}) = u''_{ij}.$$

Then U is also the total matrix ring over the skew field $a^{-1}\mathscr{D}a$ with matrix units u'_{ij} and the total matrix ring over $b^{-1}\mathscr{D}b$ with matrix units u''_{ij}. We form subalgebras

$$\begin{aligned} B &= \{Zu_{11} + \cdots + Zu_{nn}\}, \\ B_1 &= \{Zu'_{11} + \cdots + Zu'_{nn}\}, \\ B_2 &= \{Zu''_{11} + \cdots + Zu''_{nn}\} \end{aligned}$$

of U. Then the formulas

$$\sum \lambda_{ij}u_{ij} \to \sum \lambda_{ij}u'_{ij}, \quad \sum \lambda_{ij}u_{ij} \to \sum \lambda_{ij}u''_{ij} \qquad (\lambda_{ij} \in \mathbf{Z})$$

define isomorphisms of B onto B_1 and B_2 respectively, which leave fixed every element of the center Z of U. By a known theorem (see [4]) these isomorphisms can be extended to inner isomorphisms of U. Consequently,

$$x_1^{-1}u_{ij}x_1 = a^{-1}u_{ij}a, \qquad x_2^{-1}u_{ij}x_2 = b^{-1}u_{ij}b$$

for some elements $x_1, x_2 \in U$ and all $i, j = 1, \dots, n$. This implies

$$(ax_1^{-1})u_{ij} = u_{ij}(ax_1^{-1}), \quad (bx_2^{-1})u_{ij} = u_{ij}(bx_2^{-1}) \qquad (i, j = 1, \dots, n). \tag{8}$$

Let $a_1 = ax_1^{-1}$ and $b_1 = bx_2^{-1}$. By (7) we have

$$b_1^{-1}a_1b_1 = \gamma a_1^{-1}, \quad b_1^2 = \mu \qquad (\gamma, \mu \in U). \tag{9}$$

But in view of (8) the elements a_1 and b_1 commute with every matrix unit u_{ij}. Consequently, the elements γ and μ in (9) belong to the skew field $\mathscr{D}$. So the relations (9) define the algebra A, which is a crossed product of $\mathscr{D}$

with the infinite dihedral group (see [4]), while the algebra KH_ie_{i1} is the total matrix ring of order n over the algebra A of type E over K. The idempotents $e_{i1}, \dots, e_{ir}$ in (6) permute under the action of inner automorphisms of G. Therefore, the ideals on the right side of (6) are KG-isomorphic. By standard arguments (see, for example, [5]) we obtain from this that the algebra U_i is isomorphic to the total matrix ring over the ring of KG-endomorphisms of the ideal KGe_{i1}.

It is easy to see that the ring of KG-endomorphisms of KGe_{i1} is isomorphic to the algebra KH_ie_{i1}. As we have proved, this algebra is isomorphic to the total matrix ring over the algebra A of type E over K.

Finally, we obtain that every algebra U_i $(i = 1, \dots, t)$ is a total matrix ring over some algebra A_i of type E over K.

We have considered the case when H_i/F is the infinite dihedral group. If H_i/F is the infinite cyclic group, then in this degenerate case the proof can be carried out similarly (some formulas can simply be omitted). The theorem is proved.

The next proposition reduces the description of all finitely generated modules over KG to the same problem for modules over algebras of type E.

LEMMA 1. *Let an algebra B be a tensor product of the total matrix ring K_n over a field K and an algebra A with unity over K. Then every B-module M is isomorphic to a tensor product $M_1 \otimes_K M_2$, where M_1 is an irreducible K_n-module and M_2 is an A-module. B-modules M are isomorphic if and only if the corresponding A-modules M_2 are isomorphic. In particular, M is decomposable if and only if M_2 is decomposable.*

For the proof, see [2].

§2.

If K is an algebraically closed field, then it is easy to show that every algebra of type E over K is isomorphic either to the group algebra $K(a)$ of the infinite cyclic group (a) or to the group algebra KD of the infinite dihedral group $D = (a) \cdot (b)$, $b^{-1}ab = a^{-1}$, $b^2 = 1$.

We describe all the algebras of type E over the field $K = \mathbf{R}$ of real numbers. There exist three types of skew fields $\mathscr{D}$ containing $\mathbf{R}$ in their center, $(\mathscr{D} : \mathbf{R}) < \infty$:

I. $\mathscr{D}$ is the field of real numbers $\mathbf{R}$.
II. $\mathscr{D}$ is the field of complex numbers $\mathbf{C}$.
III. $\mathscr{D}$ is the skew field of quaternions $\mathbf{H}$.

Let A be an algebra of type (3): that is, a crossed product of the skew field $\mathscr{D}$ with the infinite cyclic group (a). Every automorphism of $\mathbf{R}$ is trivial and every $\mathbf{R}$-automorphism of $\mathbf{H}$ is an inner one. Replacing the element a by a suitable element δa $(\delta \in \mathscr{D})$, we obtain that in cases I and III the algebra A is isomorphic, respectively,

1. to the group algebra $A_1 = \mathbf{R}(a)$ of the infinite cyclic group (a) over the real field $\mathbf{R}$, or

2. to the group algebra $A_2 = \mathbf{H}(a)$ of the group (a) over the skew field $\mathbf{H}$ of quaternions.

If $\mathscr{D}$ is the complex field $\mathbf{C}$, then two new types of algebras arise:

3. the complex group algebra $A_3 = \mathbf{C}(a)$ of (a), and

4. the crossed group algebra A_4 of (a) over $\mathbf{C}$: $\lambda a = a\bar{\lambda}$, where $\bar{\lambda}$ is the complex conjugate of λ.

Now we determine all the algebras of type E which are crossed products of $\mathscr{D}$ with the infinite dihedral group.

Let $\mathscr{D} = \mathbf{R}$ be the real field. Then the algebra $A = (\mathscr{D}, a, b)$ of type (4) can be defined by the relations

$$a\lambda = \lambda a, \quad b\lambda = \lambda b, \quad b^{-1}ab = \gamma a^{-1}, \quad b^2 = \mu \qquad (\lambda, \delta, \mu \in \mathbf{R}). \tag{10}$$

Additional mappings of the form $b \to \delta b$, $a \to \eta a$ $(\delta, \eta \in \mathbf{R})$ in (10) reduce the algebra $A = (\mathscr{D}, a, b)$ to algebras with the relations:

5. $A_5 = (\mathbf{R}, a, b)$: $a\lambda = \lambda a$, $b\lambda = \lambda b$, $b^{-1}ab = a^{-1}$, $b^2 = 1$ $\quad(\lambda \in \mathbf{R})$,

6. $A_6 = (\mathbf{R}, a, b)$: $a\lambda = \lambda a$, $b\lambda = \lambda b$, $b^{-1}ab = -a^{-1}$, $b^2 = 1$ $\quad(\lambda \in \mathbf{R})$,

7. $A_7 = (\mathbf{R}, a, b)$: $a\lambda = \lambda a$, $b\lambda = \lambda b$, $b^{-1}ab = a^{-1}$, $b^2 = -1$ $\quad(\lambda \in \mathbf{R})$.

(The relations $b^{-1}ab = -a^{-1}$ and $b^2 = -1$ can be reduced to case 6 by the additional mapping $b \to ab$.) We remark that A_5 is the group algebra of the infinite dihedral group over the real field.

Now let $\mathscr{D}$ be the skew field $\mathbf{H}$ of quaternions. Then the algebra $A = (\mathscr{D}, a, b)$ can be defined by (10), where λ, γ, $\mu \in \mathbf{H}$. Since $\sqrt{\mu} \in \mathbf{H}$ holds for every element $\mu \in \mathbf{H}$, we have two types of algebras:

8. $A_8 = (\mathbf{H}, a, b)$: $a\lambda = \lambda a$, $b\lambda = \lambda b$, $b^{-1}ab = a^{-1}$, $b^2 = 1$ $\quad(\lambda \in \mathbf{H})$,

9. $A_9 = (\mathbf{H}, a, b)$: $a\lambda = \lambda a$, $b\lambda = \lambda b$, $b^{-1}ab = -a^{-1}$, $b^2 = 1$ $\quad(\lambda \in \mathbf{H})$,

where A_8 is the group algebra of the infinite dihedral group over $\mathbf{H}$.

Now assume that $\mathscr{D}$ is the complex field $\mathbf{C}$. If the algebra $A = (\mathscr{D}, a, b)$ is defined by (10), then it can be reduced to the form

10. $A_{10} = (\mathbf{C}, a, b)$: $a\lambda = \lambda a$, $b\lambda = \lambda b$, $b^{-1}ab = a^{-1}$, $b^2 = 1$ $\quad(\lambda \in \mathbf{C})$.

This is the group algebra of the infinite dihedral group over the complex field.

Let the algebra A be given by the relations:

$$A = (\mathbf{C}, a, b)\colon a\lambda = \lambda a, \ b\lambda = \bar{\lambda} b, \ b^{-1}ab = \gamma a^{-1}, \ b^2 = \mu, \tag{11}$$

where λ, γ, $\mu \in \mathbf{C}$. Then μ is invariant under the automorphism $\lambda \to \bar{\lambda}$ of $\mathbf{C}$, and so μ is a real number. But $a = b^{-2}ab^2 = \bar{\gamma}\gamma^{-1}a$ implies $\gamma = \bar{\gamma}$; that is, δ is also a real number. Now the mappings $b \to \delta b$, $a \to \delta a$, and $b \to ab$ reduce to the algebras with the relations:

11. $A_{11} = (\mathbf{C}, a, b)$: $a\lambda = \lambda a$, $b\lambda = \bar{\lambda} b$, $b^{-1}ab = a^{-1}$, $b^2 = 1$ $\quad(\lambda \in \mathbf{C})$,

12. $A_{12} = (\mathbf{C}, a, b)$: $a\lambda = \lambda a$, $b\lambda = \bar{\lambda} b$, $b^{-1}ab = -a^{-1}$, $b^2 = 1$ $\quad(\lambda \in \mathbf{C})$,

13. $A_{13} = (\mathbf{C}, a, b)$: $a\lambda = \lambda a$, $b\lambda = \bar{\lambda} b$, $b^{-1}ab = a^{-1}$, $b^2 = -1$ $\quad(\lambda \in \mathbf{C})$.

Finally, we consider the algebra $A = (\mathbf{C}, a, b)$ with the relations

$$A = (\mathbf{C}, a, b)\colon a\lambda = \bar{\lambda} a, \ b\lambda = \lambda b, \ b^{-1}ab = \gamma a^{-1}, \ b^2 = \mu \tag{12}$$

($\lambda, \gamma, \mu \in \mathbf{C}$). Since $\sqrt{\mu} \in \mathbf{C}$ and, as above, $\gamma \in \mathbf{R}$, the mappings $a \to \delta a$, $b \to \eta b$, and $b \to ab$ ($\delta, \eta \in \mathbf{C}$) reduce the equations (12) to the types:

14. $A_{14} = (\mathbf{C}, a, b)$: $a\lambda = \bar{\lambda} a$, $b\lambda = \lambda b$, $b^{-1}ab = a^{-1}$, $b^2 = 1$ $\quad(\lambda \in \mathbf{C})$,

15. $A_{15} = (\mathbf{C}, a, b)$: $a\lambda = \bar{\lambda} a$, $b\lambda = \lambda b$, $b^{-1}ab = -a^{-1}$, $b^2 = 1$ $\quad(\lambda \in \mathbf{C})$.

This completes the classification of algebras of type E over the real field $\mathbf{R}$. (If $a\lambda = \bar{\lambda} a$ and $b\lambda = \bar{\lambda} b$ hold in A, then by the mapping $b \to ab$ we obtain relations of type (12).)

§3.

Let R be an arbitrary ring with identity. In accordance with the terminology of general ring theory we call an element $r \in R$ *regular* if r is not a left or right zero divisor, and we call an R-module $M \neq 0$ a *torsion-free module* if $rx \neq 0$ for all $x \in M$, $x \neq 0$, and all regular elements $r \in R$.

In this section we give a full description of all finitely generated torsion-free modules over all the algebras $A = (\mathscr{D}, a, b)$ of type E over $\mathbf{R}$ with zero divisors.

First we consider the algebra $\mathscr{D}(a)$: $\lambda a = a\lambda^{\psi}$, where ψ is an $\mathbf{R}$-isomorphism of a skew field $\mathscr{D}$. Every element $x \in \mathscr{D}(a)$, $x \neq 0$, can be expressed in the form

$$x = \sum_{i=0}^{n} \lambda_{r+i} a^{r+i} \qquad (\lambda_{r+i} \in \mathscr{D},\ \lambda_r \neq 0 \text{ and } \lambda_{r+n} \neq 0,\ r, n \in \mathbf{Z},\ n \geq 0).$$

We shall call n the *norm* $N(x)$ of the element x. It is easy to see that the function $N(x)$ satisfies the following conditions.

1. If $x, y \neq 0$ then $N(x \cdot y) = N(x) + N(y)$.

2. $N(x) = 0$ if and only if $x = \lambda a^i$, $\lambda \in \mathscr{D}$, $\lambda \neq 0$. The elements λa^i with $N(\lambda a^i) = 0$ are exactly all the invertible elements in $\mathscr{D}(a)$.

3. $N(\sum_i \lambda_i a^i) = N(\sum_i \lambda_i a^{-i})$ if at least one of the coefficients λ_i is not zero.

4. The ring $\mathscr{D}(a)$ is a left and right Euclidean ring with respect to the norm $N(x)$; that is, if $x, y \in \mathscr{D}(a)$ and $y \neq 0$, then $x = q_1 y + z_1 = y q_2 + z_2$ for some q_i, $z_i \in \mathscr{D}(a)$, where either $z_i = 0$ or $N(z_i) < N(y)$ $(i = 1, 2)$.

By properties 1 and 4 of the norm $N(x)$ the ring $\mathscr{D}(a)$ contains no zero divisors and is a principal left (and right) ideal ring. It is easy to show that all the divisors of unity in $\mathscr{D}(a)$ have the form λa^i $(0 \neq \lambda \in \mathscr{D})$.

Let M be a finitely generated left $\mathscr{D}(a)$-module. Then, by known theorems (see, for example, [6]), M decomposes into the direct sum $M = N \dot{+} T$, where N is a free $\mathscr{D}(a)$-module (being a direct sum of n modules $\mathscr{D}(a)$), and T is the set of all nonregular elements in M (every element of T has a nontrivial annihilator in $\mathscr{D}(a)$). The submodule T is uniquely determined in M. The number n is an invariant of the module M and is called the *rank* of M over $\mathscr{D}(a)$.

LEMMA 2. *Let $A = (\mathscr{D}, a, b)$ be an algebra of type E. An element $r \in A$ is regular if and only if the left ideal Ar and the right ideal rA contain nonzero elements from the subalgebra $\mathscr{D}(a)$.*

PROOF. Let $r = \lambda_0 + \lambda_1 b$ be a regular element ($\lambda_i \in \mathscr{D}(a)$). If $t = \gamma_0 + \gamma_1 b$ ($\gamma_i \in \mathscr{D}(a)$), then

$$tr = (\gamma_0 + \gamma_1 b)(\lambda_0 + \lambda_1 b) = (\gamma_0\lambda_0 + \gamma_1\overline{\lambda}_1\mu) + (\gamma_0\lambda_1 + \gamma_1\overline{\lambda}_0)b, \tag{13}$$

where $b\lambda = \overline{\lambda}b$ for all $\lambda \in \mathscr{D}(a)$ and $b^2 = \mu \in \mathscr{D}$. It is clear that there exist elements $\gamma_0, \gamma_1 \in \mathscr{D}(a)$, $\gamma_0, \gamma_1 \neq 0$, such that

$$\gamma_0\lambda_1 + \gamma_1\overline{\lambda}_0 = 0. \tag{14}$$

For γ_0 and γ_1 satisfying (14) the element $\gamma_0\lambda_0 + \gamma_1\overline{\lambda}_1\mu \in \mathscr{D}(a)$ is not zero, because r is a regular element. So by (13) the ideal Ar contains a nonzero element from $\mathscr{D}(a)$. It can be proved analogously that $rA \cap \mathscr{D}(a) \neq 0$.

Conversely, let $t_1 r, rt_2 \in \mathscr{D}(a)$ and $t_1 r, rt_2 \neq 0$. Then r is a regular element, because A is a free left and free right $\mathscr{D}(a)$-module. The lemma is proved.

COROLLARY. *An A-module M is a torsion-free module if and only if M is a $\mathscr{D}(a)$-torsion-free module.*

PROOF. If M is a torsion-free module, then M is also a $\mathscr{D}(a)$-torsion-free module, because every nonzero element from $\mathscr{D}(a)$ is regular. Let M be a $\mathscr{D}(a)$-torsion-free module and $rx = 0$, where $x \in M$, $x \neq 0$, and $r \in A$ is a regular element. By Lemma 2 there exists an element $tr \in \mathscr{D}(a) \cap Ar$, $tr \neq 0$. Then $trx = 0$ implies a contradiction.

THEOREM 2. *Let A be an algebra of type E over the field K. The following propositions are equivalent.*

1) *A has zero divisors.*
2) *There exists an element $\lambda a^i b \in A$ such that $\lambda \in \mathscr{D}$ and $(\lambda a^i b)^2 = 1$.*
3) *There exists an A-module which is a free cyclic $\mathscr{D}(a)$-module.*

PROOF. We show that 2) implies 1). Indeed, if $(\lambda a^i b)^2 = 1$, then $(1 + \lambda a^i b)(1 - \lambda a^i b) = 0$ for an arbitrary field K.

We prove that 3) implies 2). Indeed, let M be an A-module, a free cyclic $\mathscr{D}(a)$-module $M = (x)$. Then $bx = \delta x$, $\delta \in \mathscr{D}(a)$, and so $b^2 x = \mu x = \delta^\varphi \delta x$ ($b\lambda = \lambda^\varphi b$ for all the $\lambda \in \mathscr{D}(a)$, and $b^2 = \mu \in \mathscr{D}$). Since (x) is a free $\mathscr{D}(a)$-module, $\mu = \delta^\varphi\delta$ and $\mu^{-1}\delta^\varphi\delta = 1$, which implies

$$\delta^{-1}(\delta^\varphi)^{-1}\mu = 1. \tag{15}$$

By (15),

$$(\delta^{-1}b)^2 = \delta^{-1}b\delta^{-1}b = \delta^{-1}(\delta^\varphi)^{-1}\mu = 1.$$

Here δ, as a divisor of unity in $\mathscr{D}(a)$, has the form $\delta = \lambda a^i$, $\lambda \in \mathscr{D}$.

1) implies 2). Let $t \in A$, $t \neq 0$, and suppose the left annihilator $\mathscr{I}$ of t in A is not zero. If the element $q = \lambda_0 + b\lambda_1$ ($\lambda_0, \lambda_1 \in \mathscr{D}(a)$) runs over the

ideal $\mathscr{I}$, then in view of $b^2 \in \mathscr{D}$ the elements λ_0 and λ_1 run over the same left ideal $\mathscr{I} = (p)$ in $\mathscr{D}(a)$. Consequently, the ideal $\mathscr{I}$ contains the element

$$p + bqp = (1 + bq)p. \tag{16}$$

Consider the element

$$t = (1 - bq)(1 + bq) = 1 - (bq)^2 \in \mathscr{D}(a). \tag{17}$$

If $t \neq 0$, then by (16) and (17) the ideal $\mathscr{I}$ contains a nonzero element from $\mathscr{D}(a)$, which contradicts the fact that A is a free left $\mathscr{D}(a)$-module. So $(bq)^2 = 1$, and then $q = \lambda a^i$, $\lambda \in \mathscr{D}$.

2) implies 3). Assume that $b_1 = \lambda a^i b$ and $b_1^2 = 1$, $\lambda \in \mathscr{D}$. Then $A(1 + b_1)$ is a left ideal of A, being a free cyclic $\mathscr{D}(a)$-module. Indeed, every element $x \in A$ can be expressed in the form $x = \lambda_0 + \lambda_1 b_1$, where λ_0, $\lambda_1 \in \mathscr{D}(a)$. Then

$$xu = (\lambda_0 + \lambda_1 b_1)(1 + b_1) = (\lambda_0 + \lambda_1)(1 + b_1) = (\lambda_0 + \lambda_1)u, \qquad \lambda_0 + \lambda_1 \in \mathscr{D}(a).$$

Here $\delta_1 u = \delta_2 u$ $(\delta_1, \delta_2 \in \mathscr{D}(a))$ clearly implies $\delta_1 = \delta_2$. So Au is a free $\mathscr{D}(a)$-module generated by u. The lemma is proved.

REMARK. If $M = (x)$ is an A-module that is a free cyclic $\mathscr{D}(a)$-module, and $b_1 x = x$, $b_1 = \lambda a^i$, and $b_1^2 = 1$, then the module M is isomorphic to the module $M(1 + b_1)$. The isomorphism θ between these modules is given by

$$\theta(\lambda x) = \lambda(1 + b_1), \quad \lambda \in \mathscr{D}(a).$$

Indeed, as follows from the proof of Theorem 2, θ is a one-to-one map of M onto $M(1 + b_1)$. We prove that θ is an A-isomorphism.

Let $b_1\delta = \overline{\delta} b_1$ for all $\delta \in \mathscr{D}(a)$. We have

$$(\lambda_0 + \lambda_1 b_1)\delta x = (\lambda_0 + \lambda_1 \delta)x, \qquad \lambda_0, \lambda_1, \delta \in \mathscr{D}(a).$$

This implies

$$\theta(\lambda_0 + \lambda_1 b_1)\delta x = (\lambda_0 + \lambda_1 \overline{\delta})(1 + b_1) = (\lambda_0 + \lambda_1 b_1)\delta(1 + b_1),$$

which was to be proved.

LEMMA 3. *Let* Char $K \neq 2$. *Then every A-module M which is a free cyclic $\mathscr{D}(a)$-module is a projective A-module.*

PROOF. As follows from Lemma 2 and the remark to it, the module M is isomorphic to $M(1 + b_1)$, where $b_1 = \lambda a^i b$ and $b_1^2 = 1$. Since the unity of A decomposes into an orthogonal sum of idempotents $1 = e_1 + e_2$, $e_1 = \frac{1}{2}(1 + b_1)$, $e_2 = \frac{1}{2}(1 - b_1)$, it follows that $M \cong Ae_1$ is a projective A-module.

LEMMA 4. *Let $A = (\mathscr{D}, a, b)$ and $b^2 = 1$. If $y = \lambda b$, $\lambda \in \mathscr{D}$, and $y^2 = 1$, then the module $A(1 + y)$ is isomorphic either to $A(1 + b)$ or to $A(1 - b)$. If* Char $K \neq 2$, *then the modules $A(1 + b)$ and $A(1 - b)$ are nonisomorphic if and only if the element b commutes with every element of $\mathscr{D}$.*

PROOF. If $\lambda \neq -1$, then

$$(1 + \lambda b)(1 + b) = (1 + \lambda)(1 + b)$$

and $(1+\lambda) \neq 0$. Then the A-isomorphism between the modules $A(1+y)$ and $A(1+b)$ can be given by

$$\mu(1+y) \to \mu(1+y)(1+b) = \mu(1+\lambda)(1+b) \qquad (\mu \in \mathscr{D}(a)),$$

because $\mu(1+\lambda)$ runs over $\mathscr{D}(a)$ if μ runs over $\mathscr{D}(a)$.

Suppose that $\lambda = -1$ but the skew field $\mathscr{D}$ contains at least one element γ not commuting with b. Then

$$(1-b)\gamma(1+b) = (\gamma - \overline{\gamma})(1+b),$$

where $\overline{\gamma} = b\gamma b^{-1}$ and $\overline{\gamma} \neq \gamma$. We obtain again that the map

$$\mu(1-b) \to \mu(1-b)\gamma(1+b) = \mu(\gamma - \overline{\gamma})(1+b) \qquad (\mu \in \mathscr{D}(a))$$

is an isomorphism of the A-modules $A(1-b)$ and $A(1+b)$.

Let $\operatorname{Char} K \neq 2$ and $b\delta = \delta b$ for all $\delta \in \mathscr{D}$. We denote by N the $\mathscr{D}(a)$-submodule of $M = A(1+b)$ generated by all the elements $(1-b)x$, $x \in M$. It is easy to see that N consists of all linear combinations

$$\sum_{i=1}^{n} \lambda_i(a^i - a^{-i})(1+b) \qquad (\lambda_i \in \mathscr{D}). \tag{18}$$

It is clear that the element $1+b$ does not belong to the set (18). Consequently, $N \subsetneq M$. On the other hand, in the case of the isomorphism $A(1-b) \cong A(1+b)$, the equation $N = M$ must hold. The lemma is proved.

REMARK. In the notation of Lemma 4, it follows from (18) that the left dimension of the factor module M/N over $\mathscr{D}$ is equal to 2 if the element b commutes with every element of $\mathscr{D}$. If b does not commute with all the elements of $\mathscr{D}$, then $M = N$.

LEMMA 5. *Let $A = (\mathscr{D}, a, b)$, $(\lambda a^i b)^2 = 1$, $\lambda \in \mathscr{D}$, and $i \equiv r \pmod 2$, where $r = 0, 1$. Then the module $A(1 + \lambda a^i b)$ is isomorphic to some module $A(1 + \mu a^r b)$, where $(\mu a^r b)^2 = 1$, $\mu \in \mathscr{D}$.*

PROOF. Let $i = 2n + r$ $(r = 0, 1)$. Then

$$a^{-n}(\lambda a^{2n+r} b)a^n = \mu a^r b, \qquad \mu \in \mathscr{D},$$

and $(\mu a^r b)^2 = 1$. We form the submodule $M_1 = A(1 + \mu a^r b)$ and set $y = a^n(1 + \mu a^r b) \in M_1$. Then

$$(\lambda a^i b)y = a^n a^{-n} \lambda a^i b a^n (1 + \mu a^r b) = a^n \mu a^r b(1 + \mu a^r b) = a^n(1 + \mu a^r b) = y.$$

By the remark to Lemma 2 this implies $M_1 = Ay \cong A(1 + \lambda a^i b)$. The lemma is proved.

LEMMA 6. *Let $A = (\mathscr{D}, a, b)$, $b^2 = 1$, be an algebra of type E with zero divisors. Then there exist not more than four nonisomorphic A-modules M that are free cyclic $\mathscr{D}(a)$-modules. If for some element $\gamma \in \mathscr{D}$ the equation $(\gamma ab)^2 = 1$ holds, then M is isomorphic to one of the four modules*

$$I_1 = A(1+b), \quad I_2 = A(1-b), \quad I_3 = A(1+\gamma ab), \quad I_4 = A(1-\gamma ab).$$

In the converse case, M is isomorphic to one of the modules I_1 or I_2. If $\operatorname{Char} K \neq 2$, *then I_1 and I_2 (I_3 and I_4 respectively) are nonisomorphic if and only if the element b (γab respectively) commutes with all the elements of $\mathscr{D}$. Both of the modules I_3 and I_4 are nonisomorphic to both I_1 and I_2 for an arbitrary field K.*

PROOF. All the assertions of the lemma, except the last one, follow from Lemmas 4 and 5. We prove the last assertion. Clearly, it is sufficient to show that the modules $I_1 = A(1+b)$ and $I_3 = A(1+\gamma ab)$ are nonisomorphic. Denote by N the $\mathscr{D}(a)$-submodule of I_3 generated by all elements $(1+b)x$, $x \in I_3$. Then N is a proper submodule of M, because elements of N have the form $(\sum \lambda_i a^i)(1+\gamma ab)$, where for every i the coefficients λ_i and λ_{-i} are not simultaneously zero. Moreover,

$$(1+b)(1+\gamma ab) = [1+(\gamma a)^{-1}](1+\gamma ab) \in N$$

and

$$(1+b)a(1+\gamma ab) = [a+a^{-1}(\gamma a)^{-1}](1+\gamma ab) \in N.$$

The first relation implies $a^{-1} \equiv \mu \pmod N$, $\mu \in \mathscr{D}$, and then the second relation gives $a \equiv \mu_1 \pmod N$, $\mu_1 \in \mathscr{D}$. So the factor module M/N as a left linear space over $\mathscr{D}$ has dimension 1. It follows that I_1 and I_3 are nonisomorphic. Indeed, denote by $\widetilde{N}$ the $\mathscr{D}(a)$-submodule of I_1 generated by all the elements $(1+b)x$, $x \in I_1$. If $\operatorname{Char} K \neq 2$, then $\widetilde{N} = I_1$. A similar result holds if $\operatorname{Char} K = 2$ and b does not commute with all the elements of $\mathscr{D}$. Let $\operatorname{Char} K = 2$ and suppose b commutes with all the elements of $\mathscr{D}$. Then $\widetilde{N}$ consists of all the sums

$$\left(\sum_{i, i \neq 0} \lambda a^i\right)(1+b),$$

where the coefficients $\lambda_i, \lambda_{-i} \in \mathscr{D}$ are not simultaneously zero and in this case the left dimension of the factor module $I_1/\widetilde{N}$ is greater than 1 (see the remark to Lemma 4). The lemma is proved.

PROPOSITION 1. *Let K be a finite field with* $\operatorname{Char} K = 2$. *Then every algebra $A = (\mathscr{D}, a, b)$ of type E contains zero divisors, and there exist exactly two A-modules which are free cyclic $\mathscr{D}(a)$-modules.*

PROOF. For every element $b_1 = a^i b \in A$ the equation $b_1^2 = \mu \in \mathscr{D}$ holds. The element b_1 defines an automorphism θ: $\theta(x) = b_1^{-1} x b_1$, $x \in \mathscr{D}$, where θ has order 2 or 1. Let θ be an automorphism of order 2; denote by $\mathscr{D}_1$ the subfield of all the elements of $\mathscr{D}$, left fixed by the action of θ. Then clearly $\mu \in \mathscr{D}_1$. By known theorems of the theory of finite fields, μ^{-1} can be expressed in the form $\mu^{-1} = \gamma \cdot \gamma^\theta$ for some $\gamma \in \mathscr{D}$. Then

$$b_1 \gamma b_1 \gamma = b_1^2 \gamma^\theta \gamma = \mu^{-1}\mu = 1.$$

If θ is the identity automorphism of $\mathscr{D}$, then we choose an element $\gamma \in \mathscr{D}$ such that $\gamma^2 = \mu^{-1}$ (this is always possible), and then $(\gamma b_1)^2 = 1$. By Lemmas 2 and 6, this completes the proof.

REMARK. Let K be an arbitrary field of characteristic 2, and π the prime subfield of K. Then by Proposition 1 the only algebras of type E in the group algebra πG (G is a group containing infinite cyclic subgroups of finite index) are free cyclic $\mathscr{D}(a)$-modules. Consequently, this holds also for the algebra KG.

Let $A = (\mathscr{D}, a, b)$ be an algebra of type E over a field K with $\operatorname{Char} K = 2$ and $b^2 = 1$. Then by Lemma 6 every A-module that is a free cyclic $\mathscr{D}(a)$-module is A-isomorphic to one of the modules $I_1 = A(1+b)$ or $I_3 = A(1+\gamma ab)$, where $(\gamma ab)^2 = 1$ and $\gamma \in \mathscr{D}$; and these modules are nonisomorphic.

We investigate the extensions of I_1 by I_1 and I_3.

LEMMA 7. *Every extension of the module I_1 by the module I_3 is decomposable. The extension of I_1 by I_1 is either decomposable or isomorphic to the module A* ($\operatorname{Char} K = 2$).

PROOF. Let M be an extension of I_1 by I_1: $M = (u_1, u_2)$, where u_2 and $u_1 + Au_2$ are generators of the A-modules $M_2 = Au_2$ and M/M_2 as free cyclic $\mathscr{D}(a)$-modules and u_1, u_2 is a basis of the $\mathscr{D}(a)$-module M. We have $bu_2 = u_2$ and $bu_1 = u_1 + \delta u_2$ ($\delta \in \mathscr{D}(a)$). Then $u_1 = b^2 u_1 = u_1 + \delta u_2 + \overline{\delta} u_2$, where $\overline{\delta} = b\delta b^{-1}$. So

$$\delta + \overline{\delta} = 0, \qquad \text{or} \qquad \overline{\delta} = \delta. \tag{19}$$

Let

$$\delta = \mu_{-s} a^{-s} + \cdots + \mu_{-1} a^{-1} + \lambda_0 + \lambda_1 a + \cdots + \lambda_r a^r, \qquad \lambda_i, \mu_j \in \mathscr{D};$$

s and r are natural numbers. Then (19) implies

$$r = s, \quad b\lambda_0 = \lambda_0 b, \quad \overline{\lambda_i a^i} = \lambda_i a^{-i} \qquad (i = 1, \ldots, r). \tag{20}$$

Let $\delta_1 = \lambda_1 a + \cdots + \lambda_r a^r$. We set $u_1' = u_1 + \delta_1 u_2$. Then by (20)

$$bu_1' = u_1 + \delta u_2 + \delta_1 u_2 = u_1 + \delta_1 u_2 + \lambda_0 u_2 = u_1' + \lambda_0 u_2. \tag{21}$$

If $\lambda_0 = 0$, then from (21) it follows that M is a direct sum $M = Au_1' \dotplus Au_2$. Let $\lambda_0 \neq 0$. Then (21) implies that M as an A-module is isomorphic to A. Indeed, A is generated by $v_1 = \lambda_0$ and $v_2 = (1+b)$. Here $Au_2 \cong I_1$, $A/Au_2 = (\lambda_0 + Av_2)$, and, by (20),

$$bv_1 = b\lambda_0 = \lambda_0 + \lambda_0(1+b) = v_1 + \lambda_0 v_2.$$

Consequently, an A-isomorphism between M and A can be given by

$$\mu_1 u_1 + \mu_2 u_2 \to \mu_1 v_1 + \mu_2 v_2 \qquad (\mu_i \in \mathscr{D}(a)).$$

Now suppose that M is an extension of I_1 by I_3: $M = (u_1, u_2)$, where u_2 and $u_1 + Au_2$ are generators of the A-modules $M_2 = Au_2 \cong I_1$ and $M/M_2 \cong I_3$

as free cyclic $\mathscr{D}(a)$-modules and u_1, u_2 is a basis of M as a $\mathscr{D}(a)$-module. Then $bu_2 = u_2$ and $(\gamma ab)u_1 = u_1 + \delta u_2$, $\delta \in \mathscr{D}(a)$. So

$$u_1 = (\gamma ab)^2 u_1 = u_1 + \delta u_2 + \gamma a \overline{\delta} u_2,$$

where $\overline{\delta} = b\delta b^{-1}$. Consequently, $\delta + \gamma a \overline{\delta} = 0$, which implies

$$\delta = \gamma a \overline{\delta}. \tag{22}$$

Let

$$\delta = \mu_{-s} a^{-s} + \cdots + \mu_{-1} a^{-1} + \mu_0 + \lambda_1 a + \cdots + \lambda_r a^r \qquad (\mu_i, \lambda_j \in \mathscr{D}).$$

Then by (22)

$$\lambda_1 a = \gamma a \overline{\mu}_0,\ \lambda_2 a^2 = \gamma a \overline{\mu}_{-1} a, \ldots, \lambda_r a^r = \gamma a \overline{\mu}_{-s} a^s. \tag{23}$$

We set

$$\delta_1 = \mu_0 + \mu_{-1} a^{-1} + \cdots + \mu_{-s} a^{-s},$$

and let $u_1' = u_1 + \delta_1 u_2$. Then, by (23),

$$\gamma ab u_1' = u_1 + \delta u_2 + \gamma a \overline{\delta}_1 u_2 = u_1 + \delta_1 u_2 = u_1'.$$

Consequently, M decomposes into the direct sum $M = Au_1' \dotplus Au_2$. The lemma is proved.

THEOREM 3. *Let $A = (\mathscr{D}, a, b)$, $b^2 = 1$, be an arbitrary algebra of type E with zero divisors over a field K, and M a finitely generated torsion-free A-module. If* Char $K \neq 2$, *then M decomposes into the direct sum of A-modules which are free cyclic $\mathscr{D}(a)$-modules and each of them is isomorphic to one of four modules, I_1, I_2, I_3, I_4 defined in Lemma* 6. *If* Char $K = 2$, *then M is a direct sum of cyclic modules each of which is isomorphic either to A or to one of the modules I_1 or I_3.*

PROOF. Clearly, M is a finitely generated $\mathscr{D}(a)$-module. Consider the direct sum decomposition $M = R \dotplus T$ as a $\mathscr{D}(a)$-module, where T is the torsion submodule of M. Then T is also an A-submodule of M. Indeed, if $x \in T$ then $\mathscr{D}(a)x$ is a finite-dimensional left linear space over $\mathscr{D}$. It is also clear that $\mathscr{D}(a)bx$ is a finite-dimensional linear space over $\mathscr{D}$, and so $bx \in T$.

Furthermore, we note that M contains A-submodules that are free cyclic $\mathscr{D}(a)$-modules. Indeed, let $x \in M$, $x \neq 0$. Consider the submodule $M_1 = Ax$. Then the A-isomorphism $M_1 \cong A/I$ holds, where I is the annihilator of the element x in A. If $I = 0$, then $M_1 \cong A$, and A contains the submodule $A(1+b)$, which is a free cyclic $\mathscr{D}(a)$-module (see Lemma 6).

Let $I \neq 0$. Then I contains no nonzero elements from $\mathscr{D}(a)$ (Ax is a $\mathscr{D}(a)$-torsion-free module), and clearly the elements 1 and b are linearly dependent over $\mathscr{D}$ modulo I. Consequently, A/I is a $\mathscr{D}(a)$-torsion-free $\mathscr{D}(a)$-module of rank 1. This implies that A/I is a free cyclic $\mathscr{D}(a)$-module.

Now we shall prove the theorem by induction on the rank r of M as a $\mathscr{D}(a)$-module. For $r = 1$, the theorem is true. Let us assume that it is true for modules of rank $r \leq n$, and prove it for an arbitrary module M of rank $n+1$. Choose an A-submodule L in M that is a free cyclic $\mathscr{D}(a)$-module (as proved, such a module exists), and consider the factor module M/L. Let T/L be the torsion submodule of M/L as a $\mathscr{D}(a)$-module. Then T is a torsion-free $\mathscr{D}(a)$-module of rank 1; that is, T is a free cyclic $\mathscr{D}(a)$-module. On the other hand, as we have proved, T/L is an A-submodule of M/L too, and so T is an A-submodule of M. It is clear that M/T is a torsion-free $\mathscr{D}(a)$-module of rank n; hence the induction hypothesis can be applied to the module M/T. First assume that $\operatorname{Char} K \neq 2$. Then by the induction hypothesis

$$M/T = M_1/T \dotplus \cdots \dotplus M_n/T,$$

where all the A-modules M_i/T $(i = 1, \dots, n)$ are free cyclic $\mathscr{D}(a)$-modules. But by Lemma 3 all such modules are projective. Consequently, the direct sum decompositions

$$M_i = S_i \dotplus T \quad \text{and} \quad M = S_1 \dotplus \cdots \dotplus S_n \dotplus T$$

hold, where, by Lemma 6, all the modules S_i and T are isomorphic to one of the modules I_1, I_2, I_3, or I_4.

Let $\operatorname{Char} K = 2$. Clearly, without loss of generality it can be assumed that $T \cong A(1+b) = I_1$. In this case, by the induction hypothesis, M/T decomposes into the direct sum of A-modules:

$$M/T = M_1/T \dotplus \cdots \dotplus M_q/T \dotplus N_1/T \dotplus \cdots \dotplus N_s/T \dotplus Q_1/T \dotplus \cdots \dotplus Q_t/T,$$

where

$$M_i/T \cong A, \quad N_i/T \cong I_1, \quad \text{and} \quad Q_j/T \cong A(1+\gamma ab) = I_3$$

(see Lemma 6). Since A is a free A-module, the direct sum decomposition

$$M_i = T \dotplus M_i', \qquad M_i' \cong A \quad (i = 1, \dots, q) \tag{24}$$

holds. On the other hand, in view of Lemma 7 the submodule T is a direct summand in the modules $Q_1, \dots, Q_t$:

$$Q_i = T \dotplus Q_i', \qquad Q_i' \cong A(1+\gamma ab) = I_3 \quad (i = 1, \dots, t). \tag{25}$$

Consider the modules N_j/T $(j = 1, \dots, s)$. By Lemma 7 every one of these modules is either decomposable (T is a direct summand) or isomorphic to A. Let

$$N_1 = T \dotplus N_1', \dots, N_m = T \dotplus N_m', \qquad N_i' \cong A(1+b) = I_1 \quad (i = 1, \dots, m), \tag{26}$$

and

$$N_{m+1} \cong A, \dots, N_s \cong A.$$

If $m = s$, then by (24)–(26) M has a decomposition

$$M = T \dotplus M_1' \dotplus \cdots \dotplus M_q' \dotplus Q_1' \dotplus \cdots \dotplus Q_t' \dotplus N_1' \dotplus \cdots \dotplus N_m'.$$

If $s = m + 1$, then we obtain

$$M = M_1' \dotplus \cdots \dotplus M_q' \dotplus Q_1' \dotplus \cdots \dotplus Q_t' \dotplus N_1' \dotplus \cdots \dotplus N_m' \dotplus N_{m+1}.$$

Finally, let $s > m + 1$. As we saw in the proof of Lemma 8, each of the modules N_i $(i = m + 1, \ldots, s)$ can be given by a special system of generators

$$N_i = (u_i, u), \qquad T = Au, \qquad N_i/T = (u_i + T),$$

where u and $u_i + T$ generate T and N_i/T as $\mathscr{D}(a)$-modules; moreover,

$$bu = u, \qquad bu_i = u_i + u \quad (i = m + 1, \ldots, s). \tag{27}$$

We set

$$u'_{m+2} = u_{m+2} + u_{m+1}, \ldots, u'_s = u_s + u_{m+1}.$$

Then, by (27),

$$bu_i' = b(u_i + u_{m+1}) = u_i + u + u_{m+1} + u = u_i' \qquad (i = m + 1, \ldots, s). \tag{28}$$

Now by (24)–(26) and (28) we have a direct decomposition of M:

$$M = M_1' \dotplus \cdots \dotplus M_q' \dotplus Q_1' \dotplus \cdots \dotplus Q_t' \dotplus N_{m+1} \dotplus Au'_{m+2} \dotplus \cdots \dotplus Au'_s,$$

where

$$N_{m+1} \cong A, \qquad Au_i' \cong I_1 = A(1 + b), \qquad M_i' \cong A, \qquad Q_j' \cong I_3.$$

The theorem is proved.

DEFINITION 3. Let $A = (\mathscr{D}, a, b)$ be an algebra of type E with zero divisors, and M a finitely generated torsion-free A-module. The decomposition of M into the direct sum of submodules

$$M = M_1 \dotplus \cdots \dotplus M_s \tag{29}$$

is called *canonical* if each of the submodules M_i either is isomorphic to the module A or is a free cyclic $\mathscr{D}(a)$-module; moreover, on the right side of (29) only one of I_1 and I_2 (similarly, I_3 and I_4) can be simultaneously found if $I_1 \not\cong I_2$ ($I_3 \not\cong I_4$ respectively). (See Lemma 6.)

We remark that by Theorem 3 every finitely generated torsion-free A-module M has a canonical decomposition, because if the modules I_1 and I_2 (I_3 and I_4) are not isomorphic (in this case $\operatorname{Char} K \neq 2$), then $I_1 \dotplus I_2 = I_3 \dotplus I_4 = A$. Indeed, by Lemmas 3 and 6

$$I_1 = Ae_1, \qquad I_2 = Ae_2, \qquad I_3 = Ae_3, \qquad I_4 = Ae_4,$$

where

$$e_1 = \tfrac{1}{2}(1 + b), \quad e_2 = \tfrac{1}{2}(1 - b), \quad e_3 = \tfrac{1}{2}(1 + \gamma ab), \quad e_4 = \tfrac{1}{2}(1 - \gamma ab)$$

($b^2 = 1$, $(\gamma ab)^2 = 1$, and $\gamma \in \mathscr{D}$).

The following theorem complements Theorem 3.

THEOREM 4. *Let $A = (\mathscr{D}, a, b)$ be an arbitrary algebra of type E with zero divisors, and let M be a finitely generated torsion-free A-module. Then any two canonical decompositions $M = M_1 \dotplus \cdots \dotplus M_s$ and $M = M_1' \dotplus \cdots \dotplus M_r'$ of M are isomorphic (that is, $r = s$ and for any given i there exists a j such that $M_i \cong M_j'$, $i, j = 1, \ldots, s$).*

PROOF. We show that the components M_i on the right side of (29) are uniquely (up to isomorphism) determined by M. Denote by N (N_i, $i = 1, \ldots, s$, respectively) the $\mathscr{D}(a)$-submodule of M (of M_i) generated by all the elements $(1+b)x$, $x \in M$ ($x \in M_i$). Clearly,

$$N = N_1 \dotplus \cdots \dotplus N_s, \qquad M/N \cong M_1/N_1 \dotplus \cdots \dotplus M_s/N_s. \tag{30}$$

If $M_i \cong A$, then

$$M_i/N_i \cong A/A(1+b) \cong A(1-b)$$

is a free cyclic $\mathscr{D}(a)$-module. If $M_i \cong I_j$, then by the remark to Lemma 4 and the proof of Lemma 6 the factor module M_i/N_i as a $\mathscr{D}(a)$-module has rank 0. Consequently, the number of modules A on the right side of (29) is an invariant of M, as the $\mathscr{D}(a)$-rank of the factor module M/N.

Now we shall show that the number of components on the right side of (29) isomorphic to $I_1 = A(1+b)$ is also entirely determined by M.

Indeed, the number of M_i isomorphic to modules I_j (for different j) equals $d - 2r$, where d is the $\mathscr{D}(a)$-rank of M and r is the $\mathscr{D}(a)$-rank of M/N. Since d and r are invariants of M, so is $d - 2r$.

Further, denote by t the left dimension over $\mathscr{D}$ of the torsion submodule of M. The number t is an invariant of M. It turns out that the numbers t and $d - 2r$ allow one to determine the number of M_i in M isomorphic to I_1. For this reason we call attention to the fact that if $M_i \cong I_3$ or I_4, the left $\mathscr{D}$-dimension of the module M_i/N_i equals 1 (see the proof of Lemma 6).

Let $\operatorname{Char} K \neq 2$. Then $M_i \cong I_1$ implies $M_i = N_i$, while by the remark to Lemma 4, if $M_i \cong I_2$ and $I_1 \not\cong I_2$, then $\dim_{\mathscr{D}} M_i/N_i = 2$. Consequently, the number of M_i in M isomorphic to I_1 equals 0 if $t \geq d - 2r$ and equals $(d - 2r) - t$ if $t < d - 2r$.

Let $\operatorname{Char} K = 2$, and suppose the element b commutes with any element of $\mathscr{D}$. Then, by the remark to Lemma 4, the left $\mathscr{D}$-dimension of M_i/N_i equals 2, where $M_i \cong I_1$. Consequently, in this case $t \geq d - 2r$ and the number of M_i in M isomorphic to I_1 equals $t - (d - 2r)$.

Finally, let $\operatorname{Char} K = 2$ and suppose the element b does not commute with any element in $\mathscr{D}$. Then by the remark to Lemma 4, in the case $M_i \cong I_1$ we have $N_i = M_i$. Consequently, $t \leq d - 2r$ and the number of M_i in M isomorphic to I_1 equals $(d - 2r) - t$.

Since the role of the module I_1 can be played by any module I_1, I_2, I_3, or I_4, this proves that the components in the canonical decomposition (29) are entirely determined by M. The theorem is proved.

REFERENCES

1. S. D. Berman and K. Buzási, *Representations of the infinite dihedral group*, Publ. Math. Debrecen **28** (1981), 173–187. (Russian)

2. ——, *On representations of a group containing an infinite cyclic subgroup of finite index*, Publ. Math. Debrecen **29** (1982), 163–170; English transl. in Amer. Math. Soc. Transl. (2) **137** (1987).

3. James E. Humphreys, *Linear algebraic groups*, Springer-Verlag, 1975.

4. I. N. Herstein, *Noncommutative rings*, Math. Assoc. Amer. and Wiley, 1968.

5. B. L. van der Waerden, *Moderne Algebra*. Vols. I, II, 2nd ed. Springer-Verlag, 1937, 1940; reprint, Stechert, New York, 1943; English transl., Ungar, New York, 1949, 1950.

6. Nathan Jacobson, *The theory of rings*, Amer. Math. Soc., Providence, R.I., 1943.

Translated by K. BUZÁSI

Amer. Math. Soc. Transl.
(2) Vol. **144**, 1989

Description of All Finite-Dimensional Real Representations of Groups Containing an Infinite Cyclic Subgroup of Finite Index

S. D. BERMAN AND K. BUZASHI [BUZÁSI]

Let G be a group containing an infinite cyclic subgroup of finite index, and KG the group algebra of G over a field K. All finitely-generated KG-modules are described in [1] in the case when K is an algebraically closed field with some restriction on its characteristic (in particular, when K has characteristic zero).

Let $\mathbf{C}$, $\mathbf{R}$, and $\mathbf{H}$ denote the complex field, the real field, and the skew field of quaternions over $\mathbf{R}$, respectively. A description of all finite-dimensional (over $\mathbf{R}$) $\mathbf{R}G$-modules is given in this paper. All the modules considered here are assumed to be left and unitary.

It is shown in [2] that the group algebra $\mathbf{R}G$ decomposes into the direct sum of two-sided ideals, each a full matrix ring over some algebra (real algebra of type E). There exist 15 types of real algebras of type E. Every such algebra is a crossed group algebra of the infinite cyclic group (a) or the infinite dihedral group $(D = (a) \cdot (b),\ b^2 = 1,\ b^{-1}ab = a^{-1})$ over $\mathbf{C}$, $\mathbf{R}$, or $\mathbf{H}$.

If L denotes $\mathbf{C}$, $\mathbf{R}$, or $\mathbf{H}$, then we shall denote by $L(a)$ the group algebra of the infinite cyclic group (a) over L.

In accordance with [2] the real algebras of type E are given as follows: $A_1 = \mathbf{R}(a)$, $A_2 = \mathbf{C}(a)$, $\mathbf{A}_3 = \mathbf{H}(a)$, and $A_4 = \mathbf{C}_1(a)$ is the crossed group algebra of the infinite cyclic group (a) over $\mathbf{C}$:

$$\lambda a = a\bar{\lambda}, \qquad \lambda \in \mathbf{C} \quad (\bar{\lambda} \text{ is the conjugate of } \lambda);$$

1980 *Mathematics Subject Classification* (1985 *Revision*). Primary 20C07, 20C32.
Translation of Publ. Math. Debrecen **31** (1984), 133–144; MR **86a**: 20007.

$A_5 = \{\mathbf{R}(a), b\}$, $A_6 = \{\mathbf{C}(a), b\}$, and $A_7 = \{\mathbf{H}(a), b\}$ are group algebras of the infinite dihedral group D over $\mathbf{R}$, $\mathbf{C}$, and $\mathbf{H}$, respectively; and

$$
\begin{array}{lllll}
A_8 = \{\mathbf{C}_1(a), b\}, & b^{-1}ab = a^{-1}, & b^2 = 1, & \lambda a = a\bar{\lambda},\ \lambda b = b\lambda & (\lambda \in \mathbf{C}); \\
A_9 = \{\mathbf{R}(a), b\}, & b^{-1}ab = a^{-1}, & b^2 = -1, & \lambda a = a\lambda,\ \lambda b = b\lambda & (\lambda \in \mathbf{R}); \\
A_{10} = \{\mathbf{R}(a), b\}, & b^{-1}ab = -a^{-1}, & b^2 = 1, & \lambda a = a\lambda,\ \lambda b = b\lambda & (\lambda \in \mathbf{R}); \\
A_{11} = \{\mathbf{C}_1(a), b\}, & b^{-1}ab = -a^{-1}, & b^2 = -1, & \lambda a = a\bar{\lambda},\ \lambda b = b\lambda & (\lambda \in \mathbf{C}); \\
A_{12} = \{\mathbf{C}(a), b\}, & b^{-1}ab = a^{-1}, & b^2 = 1, & \lambda a = a\lambda,\ \lambda b = b\bar{\lambda} & (\lambda \in \mathbf{C}); \\
A_{13} = \{\mathbf{C}(a), b\}, & b^{-1}ab = a^{-1}, & b^2 = -1, & \lambda a = a\lambda,\ \lambda b = b\bar{\lambda} & (\lambda \in \mathbf{C}); \\
A_{14} = \{\mathbf{C}(a), b\}, & b^{-1}ab = -a^{-1}, & b^2 = 1, & \lambda a = a\lambda,\ \lambda b = b\bar{\lambda} & (\lambda \in \mathbf{C}); \\
A_{15} = \{\mathbf{H}(a), b\}, & b^{-1}ab = a^{-1}, & b^2 = -1, & \lambda a = a\lambda,\ \lambda b = b\lambda & (\lambda \in \mathbf{H}).
\end{array}
$$

The algebras A_8, A_{11}, A_{12}, A_{13}, and A_{14} are crossed group algebras of the infinite dihedral group D over $\mathbf{C}$, A_9 and A_{10} are crossed group algebras of D over $\mathbf{R}$, and A_{15} is a crossed group algebra of D over $\mathbf{H}$.

The description of all finite-dimensional indecomposable $\mathbf{R}$-representations of groups G of the type considered here reduces to the same problem for the algebras $A_1, \dots, A_{15}$ (see [2]).

Below we give a full description of all finite-dimensional indecomposable A_i-modules for all the algebras A_i $(i = 1, \dots, 15)$.

Let us first consider the algebras $\mathbf{R}(a)$, $\mathbf{C}(a)$, $\mathbf{H}(a)$, and $\mathbf{C}_1(a)$. The algebras $\mathbf{R}(a)$ and $\mathbf{C}(a)$ are Euclidean rings; $\mathbf{H}(a)$ and $\mathbf{C}_1(a)$ are left and right Euclidean rings (see [2]).

Let T be a commutative principal ideal domain or a noncommutative left or right principal ideal domain. An element $p \in T$ is called *prime* if the left ideal $T \cdot p$ is maximal. Elements $c, d \in T$ are called *similar* (notation, $c \sim d$) if the quotients $T/T \cdot c$ and $T/T \cdot d$ are isomorphic as T-modules. If T is a commutative ring, then the similarity of elements c and d is equivalent to the equality $(c) = (d)$. If T is a noncommutative ring, then prime elements p_1, $p_2 \in T$ are similar if and only if their bounds are equal, that is, the maximal ideals contained in $T \cdot p_1$ and $T \cdot p_2$ are equal (see [3], Chapter 3).

Now let $T = \mathbf{R}(a)$, $\mathbf{C}(a)$, $\mathbf{C}_1(a)$, or $\mathbf{H}(a)$. We describe a full system of representatives of classes of similar elements in T.

Every prime element in $\mathbf{C}(a)$ is similar to $a - \alpha$ $(\alpha \in \mathbf{C})$, while in $\mathbf{R}(a)$ it is similar either to $a - \alpha$ $(\alpha \in \mathbf{R})$ or to the element $a^2 + \alpha a + \beta$, where $x^2 + \alpha x + \beta$ is an irreducible real polynomial.

By [3] every prime element in the ring $T = \mathbf{H}(a)$ is similar to $a - \alpha$ ($\alpha \in \mathbf{R}(i)$, where $1, i, j, k$ is the standard basis of $\mathbf{H}$ over $\mathbf{R}$). Here elements $a - \alpha_1$ and $a - \alpha_2$ $(\alpha_1, \alpha_2 \in \mathbf{R}(i))$ are similar if and only if $\alpha_2 = \bar{\alpha}_1$ (conjugate in $\mathbf{R}(i)$). Here the bound of the left ideal $T(a - \alpha_1)$ is equal to $T(a^2 + (\alpha_1 - \bar{\alpha}_1)a + |\alpha_1|^2)$.

In accordance with [3] every prime element in $\mathbf{C}_1(a)$ is similar to $a - \alpha$ $(\alpha \in \mathbf{C})$ or to the element $a^2 + \alpha$ $(\alpha \in \mathbf{C})$, where α is not a negative real number. The equalities

$$(a - \alpha)(a - \bar{\alpha}) = (a - |\alpha|)(a + |\alpha|), \tag{1'}$$

$$(a^2 + \alpha)(a^2 + \bar{\alpha}) = (a^4 + (\alpha + \bar{\alpha})a^2 + \alpha \cdot \bar{\alpha}) \tag{2'}$$

hold in $T = \mathbf{C}_1(a)$. Since the elements on the right side of (1′) and (2′) lie in the center of T, it follows from (1′) and (2′) that prime elements $a - \alpha_1$ and $a - \alpha_2$ in $\mathbf{C}_1(a)$ are similar if and only if $|\alpha_1| = |\alpha_2|$, while prime elements $a^2 + \alpha_1$ and $a^2 + \alpha_2$ in $\mathbf{C}_1(a)$ are similar if and only if either $\alpha_1 = \alpha_2$ or $\alpha_1 = \bar{\alpha}_2$. This implies that every element $a - \alpha \in \mathbf{C}_1(a)$ is similar to $a - |\alpha|$.

It follows from the formulas for the bounds of prime elements $p = a - \alpha \in \mathbf{H}(a)$ ($\alpha \in \mathbf{R}(i)$) and prime elements of the form $a - \alpha$, $a^2 + \alpha \in \mathbf{C}_1(a)$ mentioned above that the bound of the left ideal $T \cdot p^m$ for such elements $p \in T = \mathbf{H}(a)$ or $\mathbf{C}_1(a)$ is equal to (q^m), where (q) is the bound of the left ideal $T \cdot p$.

LEMMA 1. *Let T be one of the rings $\mathbf{R}(a)$, $\mathbf{C}(a)$, $\mathbf{H}(a)$, or $\mathbf{C}_1(a)$. A finite-dimensional (over $\mathbf{R}$) T-module M is indecomposable if and only if $M \cong T/T \cdot p^m$, where p is a prime element in T. The modules $T/T \cdot p_1^m$ and $T/T \cdot p_2^m$, where p_1 and p_2 are primes, are isomorphic if and only if p_1 and p_2 are similar.*

PROOF. Every indecomposable finite-dimensional (over $\mathbf{R}$) T-module M is isomorphic to the quotient module $T/T \cdot c$, where $0 \subset (c) \subset T$. Let $c = p_1 \cdots p_r$ be the decomposition of c into the product of prime elements in T. By [3] the module $T/T \cdot c$ is indecomposable if and only if the prime elements $p_1, \dots, p_r$ have the same bound (q) and the bound of $T \cdot c$ is equal to (q^r). Here the indecomposable finite-dimensional (over $\mathbf{R}$) modules $T/T \cdot c_1$ and $T/T \cdot c_2$ are isomorphic if and only if the bounds of the left ideals $T \cdot c_1$ and $T \cdot c_2$ are equal.

As has been shown above, every prime element p in $T = \mathbf{H}(a)$ or $\mathbf{C}_1(a)$ is similar to an element $a - \alpha \in \mathbf{H}(a)$ ($\alpha \in \mathbf{R}(i)$) or to $a - \alpha$, $a^2 + \alpha \in \mathbf{C}_1(a)$. By the formulas for the bounds of these elements p, the bound of the ideal $T \cdot p^m$ equals (q^m), where (q) is the bound of $T \cdot p$. This completes the proof of the lemma for $T = \mathbf{H}(a)$ or $\mathbf{C}_1(a)$. In the case $T = \mathbf{R}(a)$ or $\mathbf{C}(a)$ it is clear by the theorems on finitely-generated modules over a commutative principal ideal domain.

Let $T = \mathbf{R}(a)$, $\mathbf{C}(a)$, $\mathbf{H}(a)$, or $\mathbf{C}_1(a)$. Any algebra A_i ($i = 5, \dots, 15$) contains one of these algebras T as a subalgebra, and an $\mathbf{R}$-automorphism φ of order 2: $\varphi(x) = b^{-1}xb$, $x \in T$, acts on T. Clearly, the element $p' = \varphi(p)$ is prime for any prime element $p \in T$.

LEMMA 2. *For the algebras A_i ($5 \leq i \leq 13$, $i \neq 7$) a prime element $p \in T$ is similar to $p' = \varphi(p)$ if and only if*

1. *for A_5 and A_9: $(p) = (a \pm 1)$ or $(p) = (a - \alpha)$, where $|\alpha| = 1$, $\alpha \in \mathbf{C}$;*
2. *for A_{10}: $(p) = (a^2 + 1)$;*
3. *for A_6: $(p) = (a \pm 1)$; and for A_{12} and A_{13}: $(p) = (a - \alpha)$, where $|\alpha| = 1$, $\alpha \in \mathbf{C}$.*
4. *for A_8: $(p) = (a \pm 1)$ or $(p) = (a^2 + \alpha)$, where $|\alpha| = 1$, $\alpha \notin (-\infty, 0]$, $\alpha \in \mathbf{C}$.*

5. *for* A_{11}: $(p) = (a^2 + \alpha)$, $|\alpha| = 1$, $\alpha \notin (-\infty, 0]$, $\alpha \in \mathbf{C}$,
6. *there are no prime elements* $p \in T = \mathbf{C}(a)$ *similar to* p' *in* A_{14}.

PROOF. First of all we remark that the condition $p \sim p'$ means $T \cdot p = T \cdot p'$ for A_5, A_6, A_9, A_{10}, A_{12}, A_{13}, and A_{14}.

1. Let $A = A_5$ or A_9. Then $T = \mathbf{R}(a)$. If $p = a - \alpha$, then $p' = a^{-1} - \alpha$ and $(p) = (p')$ if and only if $\alpha = \pm 1$, that is, if $(p) = (a \pm 1)$. Let $p = a^2 + \alpha a + \beta$, where $x^2 + \alpha x + \beta$ is an irreducible real polynomial. If $(a^2 + \alpha a + \beta) = (a^{-2} + \alpha a^{-1} + \beta)$, then $1/\beta = 1$, $\alpha/\beta = \alpha$, and $(p) = (a^2 + \alpha a + 1)$.

2. Let $A = A_{10}$. Then $T = \mathbf{R}(a)$. If $p = a - \alpha$, $\alpha \in \mathbf{R}$, then $p' = -a^{-1} - \alpha$, and the equality $(a - \alpha) = (-a^{-1} - \alpha)$ is equivalent to $\alpha^{-1} = -\alpha$, which is impossible. Let $p = a^2 + \alpha a + \beta$, where $x^2 + \alpha x + \beta$ is an irreducible polynomial over $\mathbf{R}$. Then $p' = a^{-2} - \alpha a^{-1} + \beta$, and $(p) = (p')$ holds if and only if $1/\beta = \beta$ and $\alpha = -\alpha/\beta$. This implies $\beta = 1$ and $\alpha = 0$. Consequently, $p = a^2 + 1$.

3. Let $A = A_6$. If $p = a - \alpha \in \mathbf{C}(a)$, then $p' = a^{-1} - \alpha$, and $T \cdot p = T \cdot p'$ is equivalent to $\alpha = \pm 1$.

Let $A = A_{12}$ or A_{13}. If $p = a - \alpha \in \mathbf{C}(a)$, then $p' = a^{-1} - \bar{\alpha}$ and $(p) = (p')$ is equivalent to $\alpha^{-1} = \bar{\alpha}$ or $|\alpha| = 1$.

4. Let $A = A_8$. If $p = a - \alpha$, $\alpha \in \mathbf{R}$, then $p' = a^{-1} - \alpha$ and $p \sim p'$ in $\mathbf{C}_1(a)$ if and only if $\alpha^2 = 1$, that is, $\alpha = \pm 1$.

Let $p = a^2 + \alpha$, $\alpha \in \mathbf{C}$, $\alpha \notin (-\infty, 0]$. Then $p' = a^{-2} + \alpha$ and $p \sim p'$ in $\mathbf{C}_1(a)$ is equivalent to $\alpha^{-1} = \bar{\alpha}$, that is, $|\alpha| = 1$.

5. Let $A = A_{11}$. Then $T = \mathbf{C}_1(a)$. If $p = a - \alpha$, $\alpha \in \mathbf{R}$, then $p' = -a^{-1} + \alpha$, and $p \sim p'$ would imply that $\alpha = -\alpha$, which is impossible. Let $p = a^2 + \alpha$, where $\alpha \notin (-\infty, 0]$. Then $p' = a^{-2} + \bar{\alpha}$, and $p \sim p'$ is equivalent to the equality $\bar{\alpha}^{-1} = \alpha$ or $\bar{\alpha}^{-1} = \bar{\alpha}$, which implies $|\alpha| = 1$.

6. Let $A = A_{14}$. Then $T = \mathbf{C}(a)$, and it can be assumed that $p = a - \alpha$ $(\alpha \in \mathbf{C})$ and $p' = -a^{-1} + \alpha$. If $(p) = (p')$, then $\alpha = -(\bar{\alpha})^{-1}$, which is impossible. Consequently, $p \nsim p'$. The lemma is proved.

Before beginning the description of finite-dimensional (over $\mathbf{R}$) modules over the algebras A_i, we formulate some general remarks on representations of algebras.

Let A be a finite-dimensional algebra with identity over a field K, and φ an automorphism of order n on A. A crossed product $B = (A, \varphi, c)$, where c is an invertible element of A, is defined in a similar way to the crossed product of a field with an automorphism. The algebra B consists of all the linear combinations

$$u = x_1 + x_2\varphi + \cdots + x_n\varphi^{n-1}, \qquad x_i \in A,$$

where $\varphi^n = c$. By definition,

$$x_1 + x_2\varphi + \cdots + x_n\varphi^{n-1} = y_1 + y_2\varphi + \cdots + y_n\varphi^{n-1}, \qquad x_i, y_j \in A,$$

if and only if $x_i = y_i$ $(i = 1, \ldots, n)$.

The operations on elements $u \in B$ are defined in the natural way, taking into account the conditions $x\varphi = \varphi x^{\varphi}$ and $\varphi^n = c$.

Let M be an A-module. Then one can form an induced B-module.

$$M^B = M \dotplus \varphi M \dotplus \cdots \dotplus \varphi^{n-1} M, \tag{1}$$

consisting of all formal linear combinations

$$y_1 + \varphi y_2 + \cdots + \varphi^{n-1} y_n \qquad (y_i \in M,\ i = 1, \ldots, n). \tag{2}$$

Each of the subspaces $\varphi^i M$ is an A-module by virtue of the convention

$$u(\varphi^i x) = \varphi^i (u^{\varphi^i} x), \qquad u \in A.$$

The action of the operator φ on the elements (2) is defined in the natural way. If $B = (A, \varphi, c)$ and $\varphi^n = c$, then, if $\operatorname{Char} K$ does not divide n, a theorem analogous to Higman's well-known theorem in the theory of induced representations of finite groups is true.

PROPOSITION 1. *Every finite K-dimensional B-module M is a component of an induced module N^B, where N is some indecomposable A-module.*

Proposition 1 can be proved by repeating the arguments in the proof of Higman's theorem in [4] word for word.

We shall apply Proposition 1 for the description of all indecomposable finite **R**-dimensional A_i-modules, where the A_i are algebras of type E ($i = 5, \ldots, 15$). Each of these algebras is a crossed product of the algebra $T = \mathbf{R}(a)$, $\mathbf{C}(a)$, $\mathbf{H}(a)$, or $\mathbf{C}_1(a)$ with the automorphism φ of order 2: $\varphi(x) = b^{-1} x b$, $x \in T$. Therefore Proposition 1 and Lemma 1 imply the following:

LEMMA 3. *Every indecomposable finite-dimensional (over* **R***) A_i-module M is a component of the induced A_i-module*

$$M^* = M \dotplus bM, \tag{3}$$

where

$$M \cong T / T \cdot p^m$$

is the indecomposable T-module described in Lemma 1.

REMARK 1. It is easy to see that the module M^* in Lemma 3, as an A-module is isomorphic to the quotient module $A / A \cdot p^m$, and so the problem of describing all indecomposable A-modules reduces to the decomposition of the modules $A / A \cdot p^m$. Further, the summands M and bM in (3) are indecomposable T-modules, and hence the direct decomposition of $A / A \cdot p^m$ can contain no more than two terms.

We remark that the submodule bM in the sum (3) is an indecomposable T-module isomorphic to the module $T / T(p'^m)$, where $p' = b^{-1} p b$.

Now we can show a simple sufficient condition for the indecomposability of M^*.

LEMMA 4. *If $M = T/(p^m)$ and the prime elements p and p' of T are not similar, then the module M^* is indecomposable.*

PROOF. By the conditions of the lemma the module M^*, considered as a T-module, is a direct sum of the two submodules

$$M \cong T/(p^m) \text{ and } bM \cong T/(p'^m);$$

moreover, the summands M and bM as T-modules are absolutely uniquely determined by the condition $p \not\sim p'$ and Lemma 1. Consequently, any A_i-module N, $0 \subset N \subset M^*$, which is a direct summand of M^* is equal to either M or bM, and so $N = M$, which is impossible.

LEMMA 5. *Let $A = A_i$ $(5 \le i \le 15;\ i \ne 9, 13)$. If, in the notation of Lemma 1, the ideal $A \cdot p$ is a two-sided ideal of the ring A, then the module M^* (see [3]) decomposes into the direct sum of two A-submodules:*

$$M^* = \{(1+b), M\} \dotplus \{(1-b), M\} = M_1 \dotplus M_2.$$

*The submodules M_1 and M_2 are nonisomorphic if and only if $(p) = (a+1)$ or $(p) = (a-1)$ and $b\lambda = \lambda b$ for all $\lambda \in$ **R**, **C**, **H** (respectively).*

PROOF. Under the conditions of the lemma, M^* can be identified with the ring $B = A/(p^m)$ considered as an A-module. Since $1 = e_1 + e_2$, where $e_1 = \frac{1}{2}(1+b)$ and $e_2 = \frac{1}{2}(1-b)$ are idempotent, we have

$$B = B \cdot e_1 \dotplus B \cdot e_2. \tag{4}$$

The summands of (4) are indecomposable, because each $B \cdot e_i$ as a T-module is isomorphic to $T/T \cdot p^m$.

If for some $\lambda \in$ **R**, **C**, **H** (respectively) $\lambda_1 b = b\lambda$, $\lambda \ne \lambda_1$, then

$$(1-b)\lambda(1+b) = (\lambda - \lambda_1)(1+b), \qquad (\lambda - \lambda_i) \ne 0,$$

which proves the isomorphism $B(1-b) \cong B(1+b)$. Let $T \cdot p \ne (a+1)$, $(a-1)$. Then

$$(1-b)a(1+b) = (a - a^{-1})(1+b) = a^{-1}(a^2-1)(1+b) \ne 0,$$

because $(p) \ne (a+1)$, $(a-1)$ and $B(1+b) \cong T/(p^m)$. Hence $B(1-b) \cong B(1+b)$.

Finally, let $p = a+1$ or $p = a-1$ and $b\lambda = \lambda b$ for all $\lambda \in$ **R**, **C**, **H** (respectively). Then the quotient algebra $A/(p)$ is isomorphic to the group algebra $K(b)$, where $K =$ **R**, **C**, or **H** and (b) is a cyclic group of order 2. This algebra has two nonisomorphic irreducible modules corresponding to the ideals

$$K(b)(1+b) = \{\lambda(1+b)\}, \quad K(b)(1-b) = \{\lambda(1-b)\}; \qquad \lambda \in K.$$

Since (p) is the radical of the algebra $A/(p^m)$, by known theorems of the theory of Artinian algebras the number of nonisomorphic indecomposable modules in the decomposition of $A/(p^m)$ is equal to the number of its irreducible nonisomorphic modules, i.e. 2. The lemma is proved.

DEFINITION. Let $A = A_i$ $(i = 5, 6, \dots, 15)$, and let $T = \mathbf{R}(a)$, $\mathbf{C}(a)$, $\mathbf{H}(a)$, or $\mathbf{C}_1(a)$ be a subalgebra corresponding to A. We call a prime element $p \in T$ an *element of first, second* or *third kind*, respectively, if for any natural number m the module $A/A \cdot p^m$ (see (3) and Remark 1) is either indecomposable, or decomposes into the direct sum of two indecomposable modules, which are isomorphic or nonisomorphic.

THEOREM 1. *Let $A = A_i$ $(i = 5, 6, 9, 10, 12, 13, 14)$. Every prime element $p \in T$ is an element of first, second, or third kind. For the algebras A_i $(i = 5, 6, 10, 12, 14)$ the element $p \in T$ is of first kind if and only if $T \cdot p \neq T \cdot p'$. If $(p) = (p')$, then p is of second kind for A_{12} and A_{14}. For A_5, A_6, and A_{10} an element p with $(p) = (p')$ is an element of second kind if $T \cdot p \neq T(a+1)$, $T(a-1)$, and an element of third kind if $T \cdot p = T(a+1)$ or $T(a-1)$. For A_9 and A_{13} every prime element $p \in T$ is of first kind. The characterization of prime elements $p \in T$ satisfying the condition $T \cdot p \neq T \cdot p'$ or $T \cdot p = T \cdot p'$ is given in Lemma 2.*

PROOF. Since the ring T is a commutative Euclidean ring for the algebras A_i considered here, if $(p) \neq (p')$ then by Lemma 4 the module $B = A/A \cdot p^m$ is indecomposable, and so p is an element of first kind. If $T \cdot p = T \cdot p'$, then $A \cdot p^m$ is a two-sided ideal of A. In this case the module B is isomorphic to the ring $A/A \cdot p^m$ as an A-module. Now the theorem follows for the algebras A_i $(i = 5, 6, 10, 12, 14)$ from Lemma 5.

Let $A = A_9$ or A_{13} and $T \cdot p = T \cdot p'$. If $A = A_9$, then $(p) = (a \pm 1)$ or $(p) = (a^2 + \alpha a + 1)$ $(T = \mathbf{R}(a))$. It is easy to show that in the first case the algebra $A/A{\cdot}p$ is isomorphic to the field $\mathbf{C}$, while in the second it is isomorphic to the skew field $\mathbf{H}$. This implies that $B = A/A \cdot p^m$ is indecomposable into a direct sum of left ideals, because $L = A \cdot p/A \cdot p^m$ is the radical of the algebra B and A/L is a skew field.

Let $A = A_{13}$ and $(p) = (p')$, $p \in \mathbf{C}(a)$. Then, by Lemma 2, $(p) = (a - \alpha)$, $|\alpha| = 1$. In this case the algebra $A/A \cdot p$ is isomorphic to the skew field $\mathbf{H}$, and, by the same argument as in the case A_9, $A/A \cdot p^m$ is indecomposable into a direct sum of left ideals. The theorem is proved.

THEOREM 2. *Let $A = A_8$ or A_{11}. For A_8 every prime element $p \in \mathbf{C}_1(a)$ of first kind is similar to the element $a + 1$ or $a - 1$, and every element of second kind is similar to an element $a^2 + \alpha$, $|\alpha| = 1$, $\alpha \neq -1$. The rest of the prime elements $p \in \mathbf{C}_1(a)$ are of first kind. If $A = A_{11}$, then the prime elements $p \sim a^2 + \alpha$, $|\alpha| = 1$, $\alpha \neq -1$, are of second kind, and the rest of the primes $p \in \mathbf{C}_1(a)$ are of first kind.*

PROOF. Let $A = A_8$. By Lemma 4 we need to investigate only primes $p \in \mathbf{C}_1(a)$ such that $p \sim p' = b^{-1}pb$. By Lemma 2 all such primes p are similar to elements $a \pm 1$ and $a^2 + \alpha$, $|\alpha| = 1$, $\alpha \neq -1$.

First we consider the case $p = a \pm 1$. We set $N = T \cdot p^m$ and let

$$M_1 = \{N, (1 + b)\} \quad \text{and} \quad M_2 = \{N, (1 - b)\}$$

be left ideals of A generated by N and $1+b$ or N and $1-b$ respectively. We shall show that the direct decomposition

$$M^* = A/A\cdot p^m = M_1/A\cdot p^m \dotplus M_2/A\cdot p^m \tag{4'}$$

holds. In the proof we apply the condition

$$b^{-1}(A\cdot p^m)b = A\cdot p^m. \tag{5}$$

First of all we find that $M_i \subset A$ ($i = 1, 2$; proper inclusion). Indeed, every element $x(1\pm b)$, where $x \in A$, can clearly be expressed in the form $\lambda(1\pm b)$, where $\lambda \in \mathbf{C}_1(a)$. Assume that

$$(\lambda_0 + \lambda_1 b)p^m + \lambda_2(1\pm b) = 1,$$

where $\lambda_i \in T$. Then

$$\lambda_0 p^m + \lambda_2 = 1, \qquad \lambda_1 b p^m \pm \lambda_2 b = 0. \tag{6}$$

By the second equality of (6) we obtain that $\lambda_2 \in A\cdot p^m$, because $b^{-1}(A\cdot p^m)b = A\cdot p^m$, and so the first equation of (6) gives a contradiction. Now let us assume that

$$x_1(1+b) + x_2(1-b) \in A\cdot p^m. \tag{7}$$

Then, as above, it can be assumed that $x_1 = \lambda_1$ and $x_2 = \lambda_2$, where λ_1, $\lambda_2 \in C_1(a)$. (7) implies that

$$x_1 + x_2 \in A\cdot p^m \quad \text{and} \quad x_1 b - x_2 b \in A\cdot p^m;$$

and, applying (5), we obtain that $x_1 - x_2 \in A\cdot p^m$, that is, $x_1, x_2 \in A\cdot p^m$. Applying (5) again, we get $x_1(1+b) \in A\cdot p^m$ and $x_2(1-b) \in A\cdot p^m$. Consequently, (4′) is a nontrivial direct decomposition of the module M^*.

Further, let $p = a^2 + \alpha$, $|\alpha| = 1$, $\alpha \neq -1$. Then $ab(A\cdot p)ab = A\cdot p$, and so

$$ab(A\cdot p^m)ab = A\cdot p^m. \tag{8}$$

We set $\hat{N} = T\cdot p^m$ and denote by $\hat{M}_1$ and $\hat{M}_2$ the left ideals of A generated by $\hat{N}$ and $1+ab$ and by $\hat{N}$ and $1-ab$, respectively. Then the direct decomposition

$$M^* = A/A\cdot p^m = \hat{M}_1/A\cdot p^m \dotplus \hat{M}_2/A\cdot p^m \tag{9}$$

holds. Formula (9) is proved in exactly the same way as (4′) was earlier, but applying (8) instead of (5).

We show that the summands of (4′) are nonisomorphic as A-modules. We first prove this fact when $m = 1$. Since $p = a\pm 1$, in this case the module $A/A\cdot p$ is a two-dimensional space over $\mathbf{C}$, and the summands $M_i/A\cdot p$ are one-dimensional spaces. Assume that there exists an A-isomorphism $\theta\colon M_1/A\cdot p \to M_2/A\cdot p$. We set

$$e_1 = \tfrac{1}{2}(1+b), \qquad e_2 = \tfrac{1}{2}(1-b). \tag{10}$$

Then

$$\theta(e_1 + A \cdot p) = \delta e_2 + A \cdot p \qquad (\delta \in \mathbf{C}).$$

Since by (10) $e_i^2 = e_i$ and $e_1 \cdot e_2 = 0$, we have

$$\theta(e_1 + A \cdot p) = \theta[e_1(e_1 + A \cdot p)] = e_1(\delta e_2 + A \cdot p) = A \cdot p,$$

which gives a contradiction.

Now let m be an arbitrary natural number. It is easy to see that each of the modules $M_1/A \cdot p^m$ and $M_2/A \cdot p^m$ as a T-module is isomorphic to the chain T-module $T/T \cdot p^m$. It follows that each of the modules

$$M_1' = A \cdot p^{m-1}e_1 + A \cdot p^m/A \cdot p^m \quad \text{and} \quad M_2' = A \cdot p^{m-1}e_2 + A \cdot p^m/A \cdot p^m$$

is determined in $M_1/A \cdot p^m$ and $M_2/A \cdot p^m$ absolutely uniquely. Further,

$$A \cdot p^{m-1} = A \cdot p^{m-1}e_1 + A \cdot p^{m-1}e_2$$

and

$$A/A \cdot p \cong A \cdot p^{m-1}/A \cdot p^m \cong M_1' \dotplus M_2'.$$

If there exists an isomorphism

$$\theta \colon M_1/A \cdot p^m \to M_2/A \cdot p^m,$$

then $\theta(M_1') \cong \theta(M_2')$, which by (10) is in contradiction with the nonisomorphism of the summands in (4′) proved in the case $m = 1$.

Now we study the decomposition (9). By (8) we have

$$\begin{aligned}(1 + ab)a(1 - ab) &= (a - a^{-1})(1 - ab) = a^{-1}(a^2 - 1)(1 - ab) \\ &\equiv a^{-1}(-\alpha - 1)(1 - ab) \pmod{A(a^2 + \alpha)}, \qquad |\alpha| = 1,\ \alpha \neq -1.\end{aligned}$$

Hence

$$(1 + ab)A(1 - ab) = A(1 - ab),$$

and the map

$$\theta[(1 + ab) + A \cdot p^m] = (1 + ab)a(1 - ab) + A \cdot p^m$$

determines an A-isomorphism of the module $\hat{M}_1/A \cdot p^m$ onto $\hat{M}_2/A \cdot p^m$.

Let $A = A_{11}$. By Lemma 2 the primes $p \in \mathbf{C}_1(a)$ with the condition $p \sim p'$ are equal up to similarity to elements $p = a^2 + \alpha$, $|\alpha| = 1$, $\alpha \neq -1$. Then (8) holds and, repeating word for word the argument for A_8 with $p = a^2 = \alpha$, we obtain a formula of type (9) for $A = A_{11}$, where the components on the right side are isomorphic. The theorem is proved.

THEOREM 3. *Let $A = A_7$ or A_{15}. For $A = A_7$ every prime element $p = a - \alpha \in \mathbf{H}(a)$ of third kind is similar to an element $a \pm 1$. The rest of the prime elements p are of first kind. For $A = A_{15}$ the primes $p \sim a - \alpha$, $|\alpha| = 1$, are of second kind, and the rest of the prime elements are of first kind.*

PROOF. Let $A = A_7$. Then A is the group algebra of the infinite dihedral group over the skew field $\mathbf{H}$ of quaternions. Let 1, i, j, k be the standard basis of **H** over **R**. As shown above, every prime element $p \in \mathbf{H}(a)$ is similar to an element $a - \alpha$, $\alpha \in \mathbf{R}(i)$. Assume that $\alpha \neq \pm 1$. We show that in this case the quotient module $A/A \cdot p^m$ is indecomposable.

The proof will be by induction on the number m. Let $m = 1$. Clearly, an arbitrary element $x \in A$ satisfies the congruence

$$x \equiv \mu_0 + \mu_1 b \pmod{A \cdot p},$$

where μ_0, $\mu_1 \in \mathbf{H}$. If $\mu_0 = \pm\mu_1$, then $\mu_0 + \mu_1 b$ is an invertible element of the group algebra $\mathbf{H}(b)$ of the cyclic group (b) of order two over **H**. Hence it is necessary to consider the case when $\mu_0 = \pm\mu_1 \neq 0$. We have

$$a(1 \pm b) = a \pm ba^{-1} \equiv \alpha \pm b\alpha^{-1} \pmod{A \cdot p}.$$

Consequently,

$$\alpha(1 \pm b) - a(1 \pm b) \equiv \pm(\alpha - \alpha^{-1})b \pmod{A \cdot p},$$

where $(\alpha - \alpha^{-1})b$ is invertible in $\mathbf{H}(a)$ because $\alpha \neq \alpha^{-1}$. Thus, if $x \not\equiv 0 \pmod{A \cdot p}$ then $(p, x) = (1)$, that is, $A/A \cdot p$ is a simple A-module.

Let us assume we have proved that $A/A \cdot p^m$ is a chain module for all $m \leq n$, $n > 1$, and all its submodules have the form

$$A \cdot p^m, A \cdot p + A \cdot p^m/A \cdot p^m, \ldots, A \cdot p^{m-1} + A \cdot p^m/A \cdot p^m. \tag{11}$$

Consider the module $A/A \cdot p^{n+1}$. We show that it is a chain module and its submodules have the form (11) with $m = n + 1$. If $x = y \cdot p$ then, by the induction hypothesis, $(y, p^n) = A \cdot p^i$ $(0 \leq i \leq n)$, and so $(x, p^{n+1}) = A \cdot p^{i+1}$.

We assume that $x \not\equiv 0 \pmod{A \cdot p}$ and prove that $(x, p^{n+1}) = (1)$. By the induction hypothesis, for some $y \in A$, $y \cdot x = 1 + z \cdot p^n$ holds; hence the ideal $I = (x, p^{n+1})$ contains the element $1 + (\mu_0 + \mu_1 b)p^n$ and so the element

$$p^n + p^n(\mu_0 + \mu_1 b)p^n = [1 + p^n(\mu_0 + \mu_1 b)]p^n \qquad (\mu_i \in \mathbf{H}), \tag{12}$$

too, which after reduction of the element (12) $\operatorname{mod} p^{n+1}$ implies

$$[1 + p(\gamma_0 + \gamma_1 b)]p^n \in I \qquad (\gamma_i \in \mathbf{H}).$$

We shall show that

$$z = 1 + p(\gamma_0 + \gamma_1 b) \not\equiv 0 \pmod{A \cdot p}.$$

Assume that

$$1 + (a - \alpha)(\gamma_0 + \gamma_1 b) = (\lambda_0 + \lambda_1 b)(a - \alpha) \tag{13}$$

$(\gamma_0, \gamma_1 \in \mathbf{H}; \lambda_0, \lambda_1 \in \mathbf{H}(a))$. It follows from (13) that $\lambda_0 = \gamma_0$ and $1 + ((a - \alpha)\gamma_0 = \gamma_0(a - \alpha)$, which implies

$$1 = \alpha \cdot \gamma_0 - \gamma_0 \cdot \alpha \qquad (\alpha \in \mathbf{R}(i),\ \gamma_0 \in \mathbf{H}). \tag{14}$$

We show that (14) gives a contradiction. Indeed, γ_0 can be expressed in the form $\gamma_0 = \beta_1 + \beta_2 j$, where β_1, $\beta_2 \in \mathbf{R}(i)$. Then by (14) we have

$$1 = \alpha(\beta_1 + \beta_2 j) - (\beta_1 + \beta_2 j)\alpha;$$

that is,

$$1 = \alpha\beta_2 j - \beta_2 j\alpha = (\alpha\beta_2 - \beta_2\bar{\alpha})j \qquad (\alpha, \beta_2 \in \mathbf{R}(i)).$$

This equality is impossible.

Thus, the ideal $I = (x, p^{n+1})$ contains an element (12) of the form $z \cdot p^n$, $z \not\equiv 0 \pmod{A \cdot p}$. Now by the induction hypothesis $(z \cdot p^n, p^{n+1}) = A \cdot p^n$, and applying the induction hypothesis again we obtain that $(x, p^n) = (1)$, and then $I = (1)$. We have obtained that every module $A/A \cdot p^m$, $p = a - \alpha$, $\alpha \neq \pm 1$, is a chain module and, in particular, an indecomposable module.

Let $p = a \pm 1$. Then $A \cdot p^m$ is a two-sided ideal of A. The algebra $B = A/A \cdot p^n$ decomposes into the direct sum of two left ideals

$$B = B(e_1 + A \cdot p^n) \dotplus B(e_2 + A \cdot p^n), \tag{15}$$

in accordance with the decomposition $1 = e_1 + e_2$ of the identity, where $e_1 = \frac{1}{2}(1 + b)$ and $e_2 = \frac{1}{2}(1 - b)$. The summands on the right side of (15) are nonisomorphic for $A = A_7$. Let

$$A = A_{15} = \{\mathbf{H}(a), b\}, \ b^{-1}ab = a^{-1}, \ b^2 = -1, a\lambda = \lambda a, \ b\lambda = \lambda b \qquad (\lambda \in \mathbf{H}).$$

To prove the theorem it suffices to investigate only the primes $p = a - \alpha$, $|\alpha| = 1$, $\alpha \in \mathbf{R}(i)$, with the condition $p \sim p' = a - \alpha^{-1}$. As above, let $1, i, j, k$ be the standard basis of $\mathbf{H}$ over $\mathbf{R}$. Then $(jb)^2 = 1$ and

$$(jb)(A \cdot p)jb = A \cdot p, \qquad p = a^2 + \alpha, \ \alpha \in \mathbf{R}(i), \ |\alpha| = 1. \tag{16}$$

We set

$$M_1 = \{A \cdot p^m, (1 + jb)\} \quad \text{and} \quad M_2 = \{A \cdot p^m, (1 - jb)\}.$$

Then the direct decomposition

$$A/A \cdot p^m = M_1/A \cdot p^m \dotplus M_2/A \cdot p^m \tag{17}$$

can be proved with the help of (16) in exactly the same way as (4′) was proved for A_8 with the help of (5).

The summands on the right side of (17) are nonisomorphic. Indeed,

$$(1 + jb)i(1 - jb) = 2i(1 - jb).$$

This condition makes it possible to construct the A-isomorphism

$$\theta\colon M_1/A \cdot p^m \to M_2/A \cdot p^m$$

by the formula

$$\theta[(1 + jb) + A \cdot p^m] = (1 + jb)i(1 - jb) + A \cdot p^m.$$

The theorem is proved.

Now the full description of all indecomposable finite-dimensional (over $\mathbf{R}$) A-modules for $A = A_i$ $(i = 1, \ldots, 15)$ can be obtained with the help of Theorems 1, 2, and 3. We formulate the results as

THEOREM 4. *Let* $A = A_i$ $(i = 1, \dots, 15)$, *and let* $T = \mathbf{R}(a)$, $\mathbf{C}(a)$, $\mathbf{H}(a)$, *or* $\mathbf{C}_1(a)$ *be a subalgebra corresponding to* A. *Denote by* M *the indecomposable finite-dimensional* (*over* $\mathbf{R}$) A-*modules. If* $i = 1, \dots, 4$, *then the modules* M *are described in Lemma* 1. *Let* $i = 5, \dots, 15$. *Then every module* M *is isomorphic to a direct summand of the module* $A/A \cdot p^m$, *where* $p \in T$ *is a prime element. The description of the components of* $A/A \cdot p^m$ *up to isomorphism is given in Theorems* 1, 2, *and* 3. *If* $m \neq n$, *then the indecomposable modules obtained in the decomposition of* $A/A \cdot p_1^m$ *are not isomorphic to the components in the decomposition of* $A/A \cdot p_2^n$. *If* $m = n$, *then the indecomposable component* M_1 *of* $A/A \cdot p_1^n$ *is isomorphic to some indecomposable component of* $A/A \cdot p_2^n$ *if and only if one of the conditions* $p_1 \sim p_2$ *or* $p_2 \sim p_1' = b^{-1} p_1 b$ *holds.*

PROOF. By Remark 1 to Lemma 3,

$$A/(p_1^n) = T/(p_1^n) + bT/(p_1^n). \tag{18}$$

Hence, $A/(p_1^n)$ as a T-module decomposes into the sum of modules isomorphic to $T/(p_1^n)$ and $T/(p_1'^n)$; moreover, they are chain modules with composition series of length n. Since the Krull–Schmidt theorem holds for finitely-generated T-modules, it follows that if $m \neq n$ then the components of $A/(p_1^n)$ cannot be isomorphic to the components of $A/(p_2^m)$. Let $n = m$, and let some indecomposable component M_1 of $A/(p_1^n)$ be isomorphic as an A-module to the indecomposable component M_2 of $A/(p_2^n)$. Then M_1 and M_2 are isomorphic as T-modules, too, and so as T-modules they are either chain modules or direct sums of two chain modules. By Lemma 1 and (18), in both cases one of the conditions $p_1 \sim p_2$ or $p_2 \sim p_1'$ holds. If $p_1 \sim p_2$, then the direct decompositions of $A/(p_1^n)$ and $A/(p_2^n)$ are clearly isomorphic. Let $p_2 \sim p_1' = b^{-1} p_1 b$; then by Lemma 4 the module $A/(p_1^n)$ is indecomposable, because

$$A/(p_1^n) \cong A/(p_1'^n) \cong A/(p_2^n).$$

The theorem is proved.

REMARK 2. All the indecomposable finite-dimensional representations of the infinite dihedral group D over an arbitrary field K with $\operatorname{Char} K \neq 2$ are described in [5], as a special case. The propositions of this paper on the algebras A_5 and A_6 give the same results. We have included them in this paper in order to make the exposition complete.

REFERENCES

1. S. D. Berman and K. Buzási, *On representations of a group containing an infinite cyclic subgroup of finite index*, Publ. Math. Debrecen **29** (1982), 163–170; English transl. in Amer. Math. Soc. Transl. (2) **137** (1987).

2. ——, *On modules over group algebras of groups containing an infinite cyclic subgroup of finite index*, Studia Sci. Math. Hungar. **16** (1981), 455–470 (1983); English transl. in this volume.

3. Nathan Jacobson, *The theory of rings*, Amer. Math. Soc., Providence, R. I., 1943.

4. Charles W. Curtis and Irving Reiner, *Representation theory of finite groups and associative algebras*, Interscience, 1962.

5. S. D. Berman and K. Buzási, *Representations of the infinite dihedral group*, Publ. Math. Debrecen **28** (1981), 173–187. (Russian)

Translated by K. BUZÁSI

Amer. Math. Soc. Transl.
(2) Vol. **144**, 1989

Connections between Integral Characteristics of the Growth of Entire Functions and of the Distribution of their Zeros

UDC 517.53

I. S. KATS

1. In this paper we use the terminology and notation of Levin's monograph [1].

In obtaining estimates for the growth of entire functions and for the distribution of their zeros, we shall use a class (W) of real functions of a real variable. A function $\xi(r)$ $(1 \le r < +\infty)$ belongs to this class if 1) $\xi(r) \ge 0$, $\forall r \in [1, +\infty)$, 2) $\xi(r)$ is convex upward on $[1, +\infty)$, 3) the right-hand derivative $\xi^+(1)$ of $\xi(r)$ at $r = 1$ is finite, and 4) $\xi(r) = O(r)$ as $r \to +\infty$. It follows from conditions 2) and 3) that $\xi(r) \in (W)$ is absolutely continuous on $[1, +\infty)$, and has at every point $r \in [1, +\infty)$ a finite right-hand derivative $\xi^+(r)$ which is nonincreasing on $[1, +\infty)$ and coincides except perhaps on a countable set with the left-hand derivative $\xi^-(r)$ ([2], Chapter 1, §10). Then when $\xi(r) \in (W)$ it follows from 1) that $\xi^+(r) \ge 0$, $\forall r \in [1, +\infty)$, and therefore $\xi(r)$ is nondecreasing on $[1, +\infty)$; and it follows from 4) that $\xi^+(r) \to 0$ as $r \to +\infty$.

We shall frequently use the symbol $\diamond$, which, when placed between two improper integrals of nonnegative functions with respect to nondecreasing functions, means that the convergence of one of the integrals implies the convergence of the other.

The following two theorems are the main results of this paper.

THEOREM A. *Let the entire function $f(z)$ have the representation*

$$f(z) = \prod_1^{\omega} G(z/a_n; p) \qquad (\omega \le \infty), \tag{1}$$

1980 *Mathematics Subject Classification* (1985 *Revision*). Primary 30D15, 30D20.
Translation of Teor. Funktsiĭ, Funktsional. Anal. i Prilozhen. Vyp. 38 (1982), 66–73; MR **84a**: 30046.

where $G(u,p)$ is a primary factor,

$$\sum_{n=1}^{\omega} |a_n|^{-p-1} < \infty, \qquad p+1 \in \mathbf{N}, \tag{2}$$

and $\xi(r) \in (W)$. Then the convergence of the integral

$$\int_1^{+\infty} \frac{n_f(r)}{r^{p+2}} \xi(r)\, dr \tag{3}$$

implies the convergence of the integral

$$\int_1^{+\infty} \frac{\ln M_f(r)}{r^p} d(-\xi^+(r)). \tag{4}$$

THEOREM B. *Under the hypotheses of Theorem* A, *if in addition the zeros $a_n = |a_n| e^{i\alpha_n}$ of $f(z)$ are distributed in such a way that the inequalities*

$$\cos(p(\theta - \alpha_n)) \geq \eta, \quad \cos((p+1)(\theta - \alpha_n)) \leq -\eta \qquad (n = 1, 2, \dots), \tag{5}$$

are satisfied for some real numbers $\eta > 0$ and θ; then the convergence of the integral (4) *implies the convergence of the integral* (3), *and consequently*

$$\int_1^{+\infty} \frac{n_f(r)}{r^{p+2}} \xi(r)\, dr \diamond \int_1^{+\infty} \frac{\ln M_f(r)}{r^p} d(-\xi^+(r)). \tag{6}$$

Notice that the supplementary conditions in Theorem B are satisfied, in particular, when all the zeros of $f(z)$ are in an angle $\theta_1 \leq \arg z \leq \theta_2$ for which $\theta_2 - \theta_1 < \pi/(2(p+1))$.

We have generalized Theorems A and B to general entire functions of finite genus p (see §4). It appears that the first result connecting an integral characteristic of the growth of an entire function with the distribution of its zeros was obtained by Valiron in 1923. According to this result ([1], Appendix 1, §2), if $\rho(r)$ is a proximate order with nonintegral $\rho = \lim_{r\to+\infty} \rho(r) > 0$, and $f(z)$ is an entire function, then

$$\int_1^{+\infty} \frac{\ln M_f(r)}{r^{\rho(r)+1}} dr \diamond \int_1^{+\infty} \frac{n_f(r)}{r^{\rho(r)+1}} dr. \tag{7}$$

Rather recently we have generalized this theorem (see [3]): the function $r^{-\rho(r)-1}$ that appears in (7), with $\rho(r)$ a proximate order, can be replaced by a member of a considerably wider class. The generalization in Theorem A overlaps, in one direction, our result in [3] and consequently Valiron's theorem. As for Theorem B, it is a generalization in the corresponding direction only for a class of entire functions that satisfy the conditions imposed on the distribution of zeros in Theorem B. In order to "complexify" this deficiency, we will introduce Theorem C (§5).

2. In the special case when $\omega = 0$, we suppose for definiteness that $f(z) \equiv 1$ in (1). In this case Theorems A and B are trivial. Consequently it is enough to give the proof only when $\omega > 0$.

As we know ([1], Chapter I, §4, Lemma 3), under the hypotheses of Theorem A,

$$r^{-p}\ln M_f(r) < k\mu_f(r) \qquad (0 < r < +\infty) \tag{8}$$

where $k = k_p < \infty$ is a real constant, and

$$\mu_f(r) = \int_1^r \frac{n_f(t)}{t^{p+1}}\,dt + r\int_r^{+\infty}\frac{n_f(t)}{t^{p+2}}\,dt \qquad (0 < r < +\infty). \tag{9}$$

Moreover, $\mu_f(r)$ is finite for all $r \in (0, +\infty)$ ([1], Chapter I, §4, proof of Lemma 1).

By what we have said, Theorem A will be established as soon as we have proved the following lemma.

LEMMA 1. *For every* $\xi(r) \in (W)$,

$$\int_1^{+\infty}\mu_f(r)\,d(-\xi^+(r)) \diamond \int_1^{+\infty}\frac{n_f(r)}{r^{p+2}}\xi(r)\,dr. \tag{10}$$

For the proof, we shall use the following well-known auxiliary proposition (see, for example, [4], Lemma 2).

LEMMA L. *If* $\omega(x) \geq 0$ *is nondecreasing on* $[a, +\infty)$ *and* $\sigma(x) \geq 0$ *is nonincreasing on the same interval, and furthermore* $\lim_{x\to+\infty}\sigma(x) = 0$ *and at least one of* $\omega(x)$ *and* $\sigma(x)$ *is continuous on* $[a, +\infty)$, *then*

$$\int_a^{+\infty}\sigma(x)\,d\omega(x) \diamond \int_a^{+\infty}\omega(x)\,d(-\sigma(x)). \tag{11}$$

PROOF OF LEMMA 1. Notice that the function $\mu_f(r)$ defined by (9) is absolutely continuous, and that at all points of continuity of $n_f(r)$, i.e., for $r \neq |a_n|$ $(n = 1, 2, \dots)$(1)

$$(\mu_f(r))' = \int_r^{+\infty}\frac{n_f(t)}{t^{p+2}}\,dt > 0. \tag{12}$$

Consequently we have, by Lemma L, for $\xi(r) \in (W)$,

$$\int_1^{+\infty}\mu_f(r)\,d(-\xi^+(r)) \diamond \int_1^{+\infty}\left(\xi^+(r)\int_r^{+\infty}\frac{n_f(t)}{t^{p+2}}\,dt\right)dr. \tag{13}$$

Applying Lemma L again, we obtain

$$\int_1^{+\infty}\left(\xi^+(r)\int_1^{+\infty}\frac{n_f(t)}{t^{p+2}}\,dt\right)dr \diamond \int_1^{+\infty}\frac{\xi(r)n_f(r)}{r^{p+2}}\,dr. \tag{14}$$

Then (10) follows from (13) and (14). This establishes Lemma 1, and with it Theorem A.

---

(1)Moreover, this is true also for $r = |a_n|$ $(n = 1, 2, \dots)$.

3. By Lemma 1, for the proof of Theorem B it is enough to establish that under the hypotheses of this theorem there is a constant $q = q(\eta) > 0$ such that

$$r^{-p} \ln M_f(r) \geq q\mu_f(r) \qquad (0 < r < +\infty). \tag{15}$$

We first consider a single primary factor

$$G(z/a_n;p) = G(z/|a_n|e^{i\alpha_n};p). \tag{16}$$

LEMMA 2. *If the real numbers θ and α_n have the properties*

$$\cos(p(\theta - \alpha_n)) \geq \eta, \qquad \cos((p+1)(\theta - \alpha_n)) \leq -\eta \tag{17}$$

where $\eta > 0$, then

$$\operatorname{Re} \ln G\left(\frac{re^{i\theta}}{|a_n|e^{i\alpha_n}};p\right) \geq \eta r^{p+1} \int_{|a_n|}^{+\infty} \frac{dt}{t^{p+1}(t+r)} \tag{18}$$

for $r > 0$ and $|a_n| \neq 0$.

PROOF. As is well known ([5], p. 92, equations (5.1) and (5.1)$'$), if $b > 0$ and $z = re^{i\varphi}$ ($r > 0$, $\operatorname{Im}\varphi = 0$), with $\varphi \neq 0 \pmod{2\pi}$, then

$$\operatorname{Re} \ln G\left(\frac{z}{b};p\right) = \operatorname{Re}\left(z^{p+1} \int_b^{+\infty} \frac{dt}{t^{p+1}(z-t)}\right). \tag{19}$$

Since, for these values of z and $t > 0$,

$$\operatorname{Re} \frac{z^{p+1}}{z-t} = r^{p+1} \frac{r\cos p\varphi - t\cos(p+1)\varphi}{(z-t)^2}, \tag{20}$$

it follows by (19) that

$$\operatorname{Re} \ln G\left(\frac{z}{b};p\right) = r^{p+1} \int_b^{+\infty} \frac{r\cos p\varphi - t\cos(p+1)\varphi}{t^{p+1}|z-t|^2}\, dt. \tag{21}$$

By the second inequality (17), $\theta \neq \alpha_n \pmod{2\pi}$. Therefore it follows from (21) that

$$\operatorname{Re} \ln G\left(\frac{re^{i\theta}}{|a_n|e^{i\alpha_n}};p\right) = r^{p+1} \int_{|a_n|}^{+\infty} \frac{r\cos(p(\theta - \alpha_3)) - t\cos((p+1)(\theta - \alpha_n))}{t^{p+1}|z-t|^2}\, dt \tag{22}$$

and, as follows from (17),

$$\operatorname{Re} \ln G\left(\frac{re^{i\theta}}{|a_n|e^{i\alpha_n}};p\right) \geq \eta r^{p+1} \int_{|a_n|}^{+\infty} \frac{r+t}{t^{p+1}|z-t|^2}\, dt. \tag{23}$$

Hence we obtain (18).

LEMMA 3. *Under the hypotheses of Theorem* B,

$$r^{-p}\ln|f(re^{i\theta})| \geq \frac{\eta}{2}\mu_f(r) \qquad (0 < r < +\infty). \tag{24}$$

PROOF. It follows from (1) and Lemma 2 that

$$\begin{aligned}\ln|f(re^{i\theta})| &= \sum_{n=1}^{\omega}\ln\left|G\left(\frac{re^{i\theta}}{a_n};p\right)\right| = \sum_{n=1}^{\omega}\operatorname{Re}\ln G\left(\frac{re^{i\theta}}{a_n};p\right)\\ &\geq \eta r^{p+1}\sum_{n=1}^{\omega}\int_{|a_n|}^{+\infty}\frac{dt}{t^{p+1}(t+r)} = \eta r^{p+1}\int_0^{+\infty}\frac{n_f(t)\,dt}{t^{p+1}(t+r)}.\end{aligned} \tag{25}$$

Moreover,

$$\int_0^{+\infty}\frac{n_f(t)\,dt}{t^{p+1}(t+r)} = \int_0^{r}\frac{n_f(t)\,dt}{t^{p+1}(t+r)} + \int_r^{+\infty}\frac{n_f(t)\,dt}{t^{p+1}(t+r)}; \tag{26}$$

$$\int_0^{r}\frac{n_f(t)\,dt}{t^{p+1}(t+r)} \geq \frac{1}{2r}\int_0^{r}\frac{n_f(t)}{t^{p+1}}\,dt; \tag{27}$$

$$\int_r^{+\infty}\frac{n_f(t)\,dt}{t^{p+1}(t+r)} \geq \frac{1}{2}\int_r^{+\infty}\frac{n_f(t)\,dt}{t^{p+2}}. \tag{28}$$

Then (24) follows from (9) and (25)–(28). This completes the proof of the lemma.

It follows trivially from the lemma that, under the hypotheses of Theorem B, inequality (15) holds with $q = \eta/2 > 0$. This, as we have already noticed, establishes Theorem B.

4. The conclusion of Theorem A remains valid for entire functions $f(z)$ that are representable in the form

$$f(z) = z^m e^{P(z)}\prod_{n=1}^{\omega}G\left(\frac{z}{a_n};p\right), \tag{29}$$

where m is the multiplicity of the zero at the origin and

$$P(z) = \sum_{j=0}^{p} b_j z^j, \tag{30}$$

provided of course that (2) is satisfied, i.e, the theorem is valid for every entire function $f(z) \not\equiv 0$ of genus at most p.

This is evident, since by Lemma L and the properties of functions of class (W) listed in §1 every function $\xi(r) \in (W)$ satisfies $\xi(r) = O(r)$ as $r \to \infty$,

$$\int_1^{+\infty} d(-\xi^+(r)) = \xi^+(1) < \infty, \tag{31}$$

and

$$\int_1^{+\infty}\frac{\xi(r)}{r^2}\,dr \diamond \int_1^{+\infty}\frac{1}{r}\xi^+(r)\,dr \diamond \int_1^{+\infty}\ln r\,d(-\xi^+(r)). \tag{32}$$

Theorem B can also be generalized to this class of functions. Moreover, it is enough to require that the inequalities (5) are satisfied only from $n = N > 1$

onward. Notice that this generalization is (like Theorem B itself) trivial if $\omega < \infty$, i.e., if $n_f(r)$ is bounded. In fact, if $p > 0$ the convergence of (3) follows from the condition $\xi(r) = O(r)$ as $r \to +\infty$, and for $p = 0$ it follows at once from (32) and the convergence of the integral (4).(2)

Let us prove the generalization of Theorem B for $\omega = \infty$.

Under the given conditions the function (29) can be represented in the form

$$f(z) = R(z)e^{Q(z)}g(z) \tag{33}$$

where $R(z)$ is a polynomial of degree $m + N - 1$, $Q(z)$ is a polynomial of degree not exceeding p, and

$$g(z) = \prod_{n=N}^{\infty} G\left(\frac{z}{a_n}; p\right) \tag{34}$$

is an entire function that satisfies the hypotheses of Theorem B. By (33),

$$\ln M_f(r) \geq \ln|f(re^{i\theta})| = \ln|R(re^{i\theta})| + \operatorname{Re} Q(re^{i\theta}) + \ln|g(re^{i\theta})|.$$

Consequently (see Lemma 3)

$$0 \leq \ln|g(re^{i\theta})| \leq \ln M_f(r) + |Q(re^{i\theta})| + |\ln|R(re^{i\theta})||.$$

If (4) converges with $\xi(r) \in (W)$, we find from the last integral, taking account of (31) and (32), that the integral

$$\int_1^{+\infty} \frac{\ln|g(re^{i\theta})|}{r^p}\, d(-\xi^+(r))$$

converges. Hence, by Lemmas 3 and 1, the integral (3) converges with g in place of f, and consequently (3) itself converges (for large r we have $n_f(r) \leq 2n_g(r)$).

5. THEOREM C. *If $\xi(t)$ $(1 \leq t < +\infty)$ is a nonincreasing positive function such that*

$$\int_1^{+\infty} \xi(t)\, dt < \infty \tag{35}$$

and

$$\inf_{t \geq 1} \frac{\xi(2t)}{\xi(t)} > 0 \tag{36}$$

then, for every entire function $f(z)$ with zeros $a_1, a_2, \ldots$, it follows from the convergence of the integral

$$\int_1^{+\infty} \ln M_f(r)\xi(r)\, dr \tag{37}$$

that

$$\int_1^{+\infty} n_f(r)\xi(r)\, dr < \infty \quad \text{and} \quad \sum_{|a_n|>1} |a_n|\xi(|a_n|) < \infty. \tag{38}$$

(2)If $f(z)$ has at least one zero, $M_f(r)$ increases at least as fast as r; and the convergence of the integral (4) implies the convergence of the last integral in (32).

PROOF. We start from the inequality

$$\int_1^r \frac{n_f(t)}{t}\,dt \le \ln M_f(r) + K, \tag{39}$$

where $K = -\ln(|f^{(m)}(0)|/m!)$ and m is the multiplicity of the zero at the point $z = 0$; when $m = 0$ we take $K = -\ln|f(0)|$. This inequality follows trivially from an equality in [1], Chapter I, §5, proof of Theorem 5.

By (35) and (39),

$$\left(\int_1^{+\infty} \ln M_f(r)\xi(r)\,dr < \infty\right) \Rightarrow \left(\int_1^{+\infty}\left(\int_1^r \frac{n_f(t)}{t}\,dt\right)\xi(r)\,dr < \infty\right). \tag{40}$$

By Lemma L and assumption (35),

$$\int_1^{+\infty}\left(\int_1^r \frac{n_f(t)}{t}\,dt\right)\xi(r)\,dr \diamond \int_1^{+\infty}\left(\int_r^{+\infty}\xi(s)\,ds\right)\frac{n_f(r)}{r}\,dr. \tag{41}$$

Since $\xi(t)$ (> 0) is nonincreasing, for every $r \ge 1$ we have

$$\int_r^{+\infty}\xi(s)\,ds > \int_r^{2r}\xi(s)\,ds \ge r\xi(2r) > r\xi(r)K_\xi, \tag{42}$$

where $K_\xi = \inf_{t\ge 1}\xi(2t)/\xi(t) > 0$ (see (36)). Therefore

$$\int_1^{+\infty}\left(\int_r^{+\infty}\xi(s)\,ds\right)\frac{n_f(r)}{r}\,dr \ge K_\xi\int_1^{+\infty} n_f(r)\xi(r)\,dr. \tag{43}$$

By (35), Lemma L, and (42),

$$\int_1^{+\infty} n_f(r)\xi(r)\,dr \diamond \int_1^{+\infty}\left(\int_r^{+\infty}\xi(s)\,ds\right)dn_f(r)$$

$$\ge K_\xi\int_1^{+\infty} r\xi(r)\,dn_f(r) = K_\xi\sum_{|a_n|\ge 1}|a_n|\xi(|a_n|). \tag{44}$$

Since $K_\xi > 0$, we obtain the conclusion of the theorem from (40), (41), (43), and (44).

REFERENCES

1. B. Ya. Levin, *Distribution of zeros of entire functions*, GITTL, Moscow, 1956; English transl., rev. ed., Amer. Math. Soc., Providence, R.I., 1980.

2. A. Zygmund, *Trigonometric series*, vol. 1, Cambridge Univ. Press, 1959.

3. I. S. Kats, *Generalization of classical theorems on the connection between the growth of entire functions and the distribution of their zeros*, All-Union Conf. Theory of Functions of a Complex Variable, Abstracts of Reports, Fiz.-Tekhn. Inst. Nizkikh Temperatur Akad. Nauk Ukrain. SSR, Kharkov, 1971. (Russian)

4. ——, *Existence of spectral functions of generalized differential systems of second order with boundary conditions at a singular endpoint*, Mat. Sb. **68** (**110**) (1965), 174–227; English transl. in Amer. Math. Soc. Transl. (2) **62** (1967).

5. A. A. Gol′dberg and I. V. Ostrovskiĭ, *The distribution of the values of meromorphic functions*, "Nauka", Moscow, 1970. (Russian)

Translated by RALPH P. BOAS

Amer. Math. Soc. Transl.
(2) Vol. **144**, 1989

Representation of H^p Functions in a Half-Plane and Some of its Applications

UDC 513.88:517.51

YU. I. LYUBARSKIĬ

Abstract. In this article we derive a theorem about the representation of functions belonging to H_+^p, the Hardy space of the upper half-plane, and some related results about the representation of functions in the Hardy space of a strip, uniqueness conditions for classes of functions in the half-plane, and the representation of entire functions of exponential type with triangular indicator diagram.

Bibliography: 17 titles.

§1. Representation of H^p functions in a half-plane and a strip

We begin with a theorem about the representation of functions in a half-plane.

THEOREM 1. *Let $\sigma(\alpha)$ be a function of bounded variation on $[0,\infty)$ for which*

$$\inf_{t\geq 0}\left|\int_0^\infty e^{-\alpha t}\,d\sigma(\alpha)\right| > 0. \tag{1}$$

Then the operator

$$S_\sigma\colon g(z) \xrightarrow{\text{def}} \int_0^\infty g(z+i\alpha)\,d\sigma(\alpha) \tag{2}$$

is invertible on H_+^p.

PROOF. First we carry out the proof for $p=2$. By the Paley–Wiener theorem [1], functions $f\in H_+^2$ are representable in the form

$$f(z) = \frac{1}{2\pi}\int_0^\infty \phi_f(t)e^{itz}\,dt, \qquad \phi_f\in L^2(0,\infty).$$

1980 *Mathematics Subject Classification* (1985 *Revision*). Primary 30D55, 46E15.

Translation of Teor. Funktsiĭ, Funktsional. Anal. i Prilozhen. Vyp. 38 (1982), 76–84; MR **84e**:30048.

With the aid of the Fourier transform the operator S_σ can be written in the form

$$S_\sigma g(z) = \frac{1}{2\pi}\int_0^\infty \phi_g(t)\int_0^\infty e^{-\alpha t}\,d\sigma(\alpha)e^{itz}\,dt.$$

Since, under the hypotheses of the theorem, the operator defined by

$$T\colon f(z) \to \frac{1}{2\pi}\int_0^\infty \frac{\phi_f(t)}{\int_0^\infty e^{-\alpha t}\,d\sigma(\alpha)}e^{itz}\,dt$$

is bounded on H_+^2, it is obviously the inverse of S_σ. This proves the theorem for $p = 2$. For the case of arbitrary $p \in (1, \infty)$ it is sufficient to extend the operator T on the set $H_+^2 \cap H_+^p$ to a bounded operator on H_+^p; we then obtain a bounded operator which is the inverse of S_σ on the dense set $H_+^2 \cap H_+^p$. In other words, it is sufficient to show that the function $m(t) \stackrel{\text{def}}{=} (\int_0^\infty e^{-\alpha t}\,d\sigma(\alpha))^{-1}$ is a multiplier on L^p for $\infty > p > 1$ (for the definition, see, for example, [2] Chapter IV, §3.1). It follows from a theorem of Marcinkiewicz ([2], Chapter IV, §6.1, Theorem 6) that in order for this function to be a multiplier it is sufficient that $\sup_{t\geq 0}|m(t)| < \infty$ and $m'(t) \in L^1(0, \infty)$. However, these properties of $m(t)$ follow directly from condition (1) and the fact that $\sigma(\alpha)$ is of bounded variation. This finishes the proof of the theorem.

Let $a < b$. We recall that $H^p(a, b)$ denotes the space of functions $f(z)$ which are holomorphic in the strip $a < \operatorname{Im} z < b$ and for which

$$\|f\|_{H^p(a,b)}^p \stackrel{\text{def}}{=} \sup_{a<y<b}\int_{-\infty}^\infty |f(x+iy)|^p\,dx < \infty.$$

Such spaces were studied in [4]. In particular, it is known that functions $f \in H^p(a, b)$ have limiting values $f(t + ia)$ and $f(t + ib)$ (in the L^p sense), and that

$$\|f\|_{H^p(a,b)} \asymp \max(\|f(t+ia)\|_{L^p}, \|f(t+ib)\|_{L^p}).$$

For the application of Theorem 1 to functions in $H^p(a, b)$ we shall need the following fact.

LEMMA 1. *Every function $f(z) \in H^p(a, b)$ $(1 < p < \infty)$ can be represented in a unique way as*

$$f(z) = g_+(z - ia) - g_-(z - ib), \tag{3}$$

where the functions $g_\pm$ belong to the Hardy spaces $H_\pm^p$ in the upper and lower half-planes, respectively.

The proof is obtained at once from the Cauchy integral formula:

$$f(z) = \frac{1}{2\pi i}\left(\int_{-\infty}^\infty \frac{f(t+ia)}{t-z+ia}\,dt - \int_{-\infty}^\infty \frac{f(t+ib)}{t-z+ib}\,dt\right),$$

where g_+ and g_- are given by the first and second terms of the right-hand side, respectively. The uniqueness of the representation (3) follows immediately from Liouville's theorem.

Lemma 1 together with Theorem 1 gives us

THEOREM 2. *Let the numbers* $\theta_1, \theta_2 > 0$ *and the function* $\sigma(\alpha)$ *of bounded variation on* $[-\theta_1, \theta_2]$ *be such that*

$$\inf_{t\geq 0}\left|\int_0^{\theta_1+\theta_2} e^{-\beta t}\, d\sigma(\beta-\theta_1)\right| > 0 \quad \text{and} \quad \inf_{t\geq 0}\left|\int_0^{\theta_1+\theta_2} e^{-\beta t}\, d\sigma(\theta_2-\beta)\right| > 0. \tag{4}$$

Then the operator $S_\sigma\colon H^p(a-\theta_1, b+\theta_2) \to H^p(a,b)$ *defined by*

$$S_\sigma\colon g(z) \to \int_{-\theta_1}^{\theta_2} g(z+i\alpha)\, d\sigma(\alpha) \tag{5}$$

is invertible.

We single out the following special case of this theorem which is of interest to us.

COROLLARY 1. *If the numbers* $A_1, A_2 \in \mathbf{C}$ *are such that for* $t \geq 0$

$$A_1 e^{\theta_1 t} + A_2 e^{-\theta_2 t} \neq 0 \quad \text{and} \quad A_2 e^{\theta_1 t} + A_1 e^{-\theta_2 t} \neq 0,$$

then every function $f(z) \in H^p(a,b)$ *permits a unique representation*

$$f(z) = A_1 g(z - i\theta_1) + A_2 g(z + i\theta_2) \tag{6}$$

with a function $g \in H^p(a-\theta_1, b+\theta_2)$*; and, moreover,*

$$\|f\|_{H^p(a,b)} \asymp \|g\|_{H^p(a-\theta_1, b+\theta_2)}. \tag{7}$$

We denote by $H^p(a,b;e^{hx})$ the space of functions $f(z)$ analytic in $a < \operatorname{Im} z < b$ and such that

$$\|f\|_{H^p(a,b;e^{hx})} \overset{\text{def}}{=} \sup_{a<y<b} \int_{-\infty}^{\infty} |f(x+iy)|^p e^{hx}\, dx < \infty.$$

COROLLARY 2. *Let* $\theta \in (0, \pi/2)$, $\theta' = \frac{\pi}{2} - \theta$, *and* $h \in \mathbf{R}$. *Every function* $f \in H^p(-\theta', \theta'; e^{hx})$ *has a unique representation of the form*

$$f(z) = g(z+i\theta) - g(z-i\theta), \tag{8}$$

where $g \in H^p(-\pi, \pi; e^{hx})$ *and*

$$\|g\|_{H^p(-\pi,\pi;e^{hx})} \asymp \|f\|_{H^p(-\theta,\theta;e^{hx})}.$$

This follows from Corollary 1 and the fact that the mapping $f(z) \to f(z)e^{zh/p}$ is an isomorphism between the spaces $H^p(a,b;e^{hx})$ and $H^p(a,b)$.

§2. Uniqueness theorems for holomorphic functions in a half-plane

We denote by $\mathbf{H}^p(\alpha;\Delta)$ the space of functions analytic in the sector $|\arg w| < \alpha$ with the norm[1]

$$\|F\|_{\mathbf{H}^p(\alpha;\Delta)} \overset{\text{def}}{=} \sup_{|\theta|<\alpha} \int_0^{\infty} |F(\rho e^{i\theta})|^p \rho^\Delta\, d\rho < \infty.$$

[1] These spaces are not new. For $p = 2$ they were considered for the first time in [5] and were investigated in detail in [6]. For arbitrary $p \in (1,\infty)$ they were introduced in [3] and [8], where it was shown, in particular, that $\mathbf{H}^p(\pi/2, 0)$ coincides with the ordinary Hardy space for the right half-plane. The notation of [5]–[8] differs somewhat from ours.

It is well known ([9]) that in order for a sequence of points $\{w_k\}_1^\infty$, $\operatorname{Re} w_k > 0$, to be the sequence of zeros of some function in H_r^p, the Hardy space of the right half-plane (and in general of a function of bounded type in the right half-plane), it is necessary and sufficient that $\sum_k (1+|w_k|^2)^{-1}\operatorname{Re} w_k < \infty$.

We pick a number $\theta \in (0, \frac{\pi}{2})$ and set $\theta' = \frac{\pi}{2} - \theta$. One can compare with each sequence of points $\{w_k\}_1^\infty$ lying in the sector $|\arg w| < \theta'$, the sequence of functionals L_k defined on H_r^p by

$$L_k[F] \overset{\text{def}}{=} F(w_k e^{i\theta}) - F(w_k e^{-i\theta}). \tag{9}$$

When does there exist a function in H_r^p for which these functionals vanish? The answer to this question is given by the following theorem.

THEOREM 3. *Let the sequence $\{w_k\}_1^\infty$ be such that $|\arg w_k| < \theta'$, and let the functionals L_k be defined by* (9). *In order that there exist a function $F \in H_r^p$ for which*

$$L_k[F] = 0, \qquad k = 1, 2, \ldots, \tag{10}$$

it is necessary and sufficient that

$$\sum_k (1 + |w_k|^{\pi/\theta'})^{-1} \operatorname{Re} w_k^{\pi/(2\theta')} < \infty. \tag{11}$$

PROOF. Let the function $F \in H_r^p$ satisfy (10). Then $\{w_k\}_1^\infty$ is the sequence of zeros of a function

$$G(w) = F(e^{i\theta}w) - F(e^{-i\theta}), \tag{12}$$

which belongs to $\mathbf{H}^p(\theta', 0)$. The mapping $G(w) \mapsto \Gamma(w) \overset{\text{def}}{=} G(w^{2\theta'/\pi})$ establishes an isomorphism between $\mathbf{H}^p(\theta'; 0)$ and $\mathbf{H}^p(\frac{\pi}{2}; \frac{2\theta'}{\pi} - 1)$. Therefore $\{w_k^{\pi/(2\theta')}\}_{k=1}^\infty$ is the sequence of zeros of the function $\Gamma(w) \in \mathbf{H}^p(\frac{\pi}{2}; \frac{2\theta'}{\pi} - 1)$, which is a function of bounded type. Consequently (11) is satisfied. Conversely,it follows from (11) that there exists a function $\Gamma(w) \in \mathbf{H}^p(\frac{\pi}{2}; \frac{2\theta'}{\pi} - 1)$ which vanishes at the points $\{w_k^{\pi/(2\theta)}\}$, and therefore also a function $G \in \mathbf{H}^p(\theta', 0)$ which vanishes at the points w_k.(2) To finish the proof of the theorem it is sufficient to show that the function $G \in \mathbf{H}^p(\theta'; 0)$ has a representation of the form (12), where $F \in H_r^p$. This representation is obtained from Corollary 2 of Theorem 2 and from the fact that the mapping $F(w) \to f(z) \overset{\text{def}}{=} F(e^z)$ establishes an isomorphism between $\mathbf{H}^p(\alpha; 0)$ and $H^p(-\alpha, \alpha; e^x)$ (defined in §1). This finishes the proof of the theorem.

Analogously one obtains a theorem about interpolation for a sequence of functionals $\{L_k\}$.

(2)The results about interpolation and uniqueness of functions in angular domains (and more general results which allow for multiple nodes) used in this and some of the following theorems and the assertions about completeness and the basis property which are related to these results were obtained in [7].

THEOREM 4. *Let the sequence* $\{w_k\}_1^\infty$ *lie in the sector* $|\arg w| < \theta'$ *and satisfy condition* (11). *Let the functionals* L_k *be defined by* (9). *In order that for every sequence* $\{a_k\} \in l^p$ *there exist a function* $F \in H_r^p$ *such that* $L_k[F] = a_k|w_k|^{(\pi/(2\theta')-1)/2}$, *it is necessary and sufficient that*

$$0 < \inf \operatorname{Re} w_k^{\pi/(2\theta')} \leq \sup \operatorname{Re} w_k^{\pi/(2\theta')} < \infty \quad \text{and} \quad \inf_{k \neq j} |w_k^{\pi/(2\theta')} - w_j^{\pi/(2\theta')}| > 0. \tag{13}$$

Results about interpolation in a half-plane which were obtained in [10] are used in the proof of this theorem.

We mention the theorems about completeness and the basis property which are dual, respectively, to Theorems 3 and 4 about uniqueness and interpolation.

THEOREM 3′. *Let the sequence of points* $\{w_k\}_1^\infty$ *lie in the sector* $|\arg w| < \theta'$. *A necessary and sufficient condition for the system of functions*

$$\{(w + \overline{w}_k e^{-i\theta})^{-1} + (w + \overline{w}_k e^{i\theta})^{-1}\}_{k=1}^\infty \tag{14}$$

to be incomplete in H_+^p *is that they satisfy condition* (11).

We mention the following corollary, which is obtained from the case $p = 2$ of this theorem with the help of the Fourier transform.

COROLLARY. *Under the hypotheses of Theorem* 3, *a necessary and sufficient condition for the system*

$$\{e^{-w_k e^{i\theta} t} - e^{-\overline{w}_k e^{-i\theta} t}\}_{k=1}^\infty \tag{14′}$$

to be incomplete in $L^2(0, \infty)$ *is that* (11) *be satisfied.*([3])

THEOREM 4′. *Under the hypotheses of Theorem* 3, *a necessary and sufficient condition for the system* (14) *to be a Riesz basis* (*after normalization*) *in the closure of its linear span is that* $\{w_k\}$ *satisfy the conditions* (13).

Theorems 3 and 4, and 3′ and 4′, respectively, permit natural generalizations, analogous to the generalization of Theorem 1, with replacement of the linear combinations of values of the function F by an integral with respect to a suitable function of bounded variation. For the sake of brevity we will not give the complete formulation here.

§3. Decomposition of entire functions of exponential type

In this section we study the case $p = 2$. Let D be the triangle with vertices at the points 0, $-ae^{-i\theta}$, and $-ae^{i\theta}$ with $a > 0$ and $\theta \in (0, \pi/2)$, let $I = [-a, 0]$ be a segment of the real axis, and let $h_D(\phi)$ and $h_I(\phi)$ be the support functions

([3])In the particular case that $w_k = k$ the system (14′) is the well-known system of A. G. Kostyuchenko, which was investigated for $L^p(0, l)$ in [11] and [16].

of these sets. We will consider the space L_D^2, which consists of all entire functions of exponential type (e.f.e.t.) with the norm

$$\|F\|_{L_D^2}^2 = \sup_{0<\phi<2\pi} \int_0^\infty |F(\rho e^{i\phi})|^2 e^{-2\rho h_D(\phi)}\, d\rho < \infty, \tag{15}$$

and the analogously defined space L_I^2.(4) By the Paley–Wiener theorem, L_I^2 coincides with the space of functions of the form $G(w) = \int_0^a e^{-wt}\psi(t)\,dt$, $\psi \in L^2(0,a)$. If $G \in L_I^2$, then obviously

$$F(w) = G(we^{i\theta}) - G(we^{-i\theta}) \tag{16}$$

belongs to L_D^2. How can one describe all functions $F \in L_D^2$ which permit such a representation? The following theorem answers this question in a particular case.

THEOREM 5. *Let $\theta = \pi/n$, where $n \geq 3$ is a natural number. In order that the e.f.e.t. $F(w) = \sum_{k=0}^\infty c_F(k)z^k \in L_D^2$ permit a representation* (16) *in terms of a function $G \in L_I^2$, it is necessary and sufficient that*

$$c_F(nl) = 0, \quad l = 1, 2, \dots. \tag{17}$$

PROOF. The necessity of condition (17) is obvious. We prove the sufficiency. The entire function $f(z) \stackrel{\text{def}}{=} F(e^z)$ of period $2\pi i$ is representable in the form $f(z) = \sum_{k=0}^\infty c_F(k)e^{kz}$. In addition, when considered as a function on the strip $|\operatorname{Im} z| < \theta'$ (here $\theta' = \frac{\pi}{2} - \theta$), $f(z)$ belongs to $H^2(-\theta', \theta'; e^x)$, and consequently is representable in the form

$$f(z) = g(z + i\theta) - g(z - i\theta), \tag{18}$$

where $g(z) \in H^2(-\frac{\pi}{2}, \frac{\pi}{2}; e^x)$. The function $g(z)$ can be extended to an entire function. Indeed, (18) can be rewritten in the form

$$g(z + 2i\theta) = f(z + i\theta) + g(z). \tag{18$'$}$$

The right-hand side in this equation is analytic for $-\frac{\pi}{2} < \operatorname{Im} z < \frac{\pi}{2}$, so that $g(z)$ is analytic for $-\frac{\pi}{2} < \operatorname{Im} z < \frac{\pi}{2} + 2\theta$. By repeating this argument we establish the analyticity of $g(z)$ in the upper half-plane, and, analogously, in the lower half-plane. From (18′) and the hypotheses of the theorem it follows that for the function $g(z)$ of period $2\pi i$

$$\begin{aligned} g(z + 2\pi i) - g(z) &= \sum_{j=0}^{n-1} [g(z + 2i(j+1)\theta) - g(z + 2ij\theta)] \\ &= \sum_{j=0}^{n-1} f(z + i(2j+1)\theta) = \sum_{j=1}^{n-1} \sum_{k \not\equiv 0 \bmod n} c_F(k)e^{kz + i(2j+1)k\theta} \\ &= \sum_{k \not\equiv 0 \bmod n} c_F(k)e^{kz + ik\theta} \sum_{j=0}^{n-1} e^{i2kj} = 0. \end{aligned}$$

(4)The space L_D^2 for an arbitrary convex polygon D was introduced and studied in [1] and [17].

Therefore $G(w) = g(\ln z)$ is analytic in the w-plane with the possible exception of the point $w = 0$, and for all values $w \neq 0$ the representation (16) holds. It remains to show that the singularity at 0 is removable and that the entire function $G(w)$ belongs to L_I^2. Both of these assertions follow from the fact that(5)

$$\sup_\phi \int_0^\infty |G(\rho e^{i\phi})|^2 e^{-2\rho h_I(\phi)}\, d\rho < \infty. \tag{19}$$

We have

$$h_I(\phi) = \begin{cases} 0, & -\pi/2 \leq \phi \leq \pi/2, \\ -a\cos\phi, & \pi/2 \leq |\phi| \leq \pi, \end{cases} \tag{20}$$

and

$$h_D(\phi) = \begin{cases} 0, & |\phi| \leq \theta', \\ -a\cos(\theta+\phi), & \theta' \leq \phi \leq \pi, \\ -a\cos(\theta-\phi), & -\pi \leq \phi \leq -\theta'. \end{cases} \tag{21}$$

Therefore, after the change of variable $\rho = e^x$, (19) becomes

$$\int_{-\infty}^{\infty} |g(x+iy)|^2 e^x\, dx < \infty, \qquad |y| < \tfrac{\pi}{2}; \tag{22A}$$

$$\int_{-\infty}^{\infty} |g(x+iy)|^2 e^{2a\cos y e^x} e^x\, dx < \infty, \qquad \tfrac{\pi}{2} \leq |y| \leq \pi. \tag{22B}$$

The first of these inequalities follows from the fact that $g \in H^2(-\frac{\pi}{2}, \frac{\pi}{2}; e^x)$. We prove the second. Suppose that $y \in (\frac{\pi}{2}, \min(\frac{\pi}{2} + 2\theta, \pi))$. In this case $y - \theta \geq \theta'$, and by virtue of (15) and (21)

$$\int_{-\infty}^{\infty} |f(x + i(y-\theta))|^2 e^{2a\cos y e^x} e^x\, dx < \infty. \tag{23}$$

If we use (18′) in the form $g(z) = f(z - i\theta) + g(z - 2i\theta)$ and take into account (22A) and (23), the required inequality for $y \in (\frac{\pi}{2}, \min(\pi, \frac{\pi}{2} + 2\theta))$ follows. If $\frac{\pi}{2} + 2\theta \geq \pi$, the proof is finished (the argument for $y \in (-\pi, -\frac{\pi}{2})$ is analogous). If, however, $\frac{\pi}{2} + 2\theta < \pi$, it is necessary to repeat these arguments for values $y \in (\frac{\pi}{2} + 2\theta, \min(\frac{\pi}{2} + 4\theta, \pi))$, employing (22B), which has already been proved for the interval $(\frac{\pi}{2}, \frac{\pi}{2} + 2\theta)$, instead of (22A), and the inequality $-\cos y > -\cos(y - 2\theta)$, which is valid for $y \in (\frac{\pi}{2} + 2\theta, \min(\frac{\pi}{2} + 4\theta), \pi))$. This process is continued until the angle $(\frac{\pi}{2}, \pi)$ is exhausted, and analogously for the angle $(-\pi, -\frac{\pi}{2})$. This finishes the proof of the theorem.

References

1. B. Ya. Levin, *Distribution of the zeros of entire functions*, GITTL, Moscow, 1956; English transl., rev. ed., Amer. Math. Soc., Providence, R.I., 1980.

2. Elias M. Stein, *Singular integrals and differentiability properties of functions*, Princeton Univ. Press, Princeton, N.J., 1970.

(5)In this case the removability of the singularity at 0 follows from the integrability of $|G(w)^2|$ with respect to area in a neighborhood of 0.

3. A. M. Sedletskiĭ, *An equivalent definition of H^p spaces in the half-plane and some applications*, Mat. Sb. **96 (138)** (1975), 75–82; English transl. in Math. USSR Sb. **25** (1975).

4. E. C. Titchmarsh, *Introduction to the theory of Fourier integrals*, Clarendon Press, Oxford, 1937.

5. M. M. Dzhrbashyan and A. E. Avetisyan, *Integral representation of a class of functions analytic in an angular domain*, Dokl. Akad. Nauk SSSR **120** (1958), 457–460; Sibirsk. Mat. Zh. **1** (1960), 383–426. (Russian)

6. M. M. Dzhrbashyan, *Integral transforms and representations of functions in the complex domain*, "Nauka", Moscow, 1966. (Russian)

7. V. M. Martirosyan, *Closure and the basis property of some biorthogonal systems and the solution of a multiple interpolation problem in angular domains*, Izv. Akad. Nauk Armyan. SSR Ser. Mat. **13** (1978), 490–531. (Russian)

8. S. A. Akopyan, *A two constants theorem for functions of class H_p*, Izv. Akad. Nauk Armyan. SSSR Ser. Mat. **2** (1967), 123–127. (Russian)

9. Kenneth Hoffman, *Banach spaces of analytic functions*, Prentice-Hall, Englewood Cliffs, N.J., 1962.

10. N. K. Nikol′skiĭ, *Treatise on the shift operator*, "Nauka", Moscow, 1980; English transl., Springer-Verlag, 1986.

11. B. Ya. Levin, *Entire functions*, Izdat. Moskov. Gos. Univ., Moscow, 1971. (Russian)

12. M. G. Dzhavadov, *On the completeness of a certain set of eigenfunctions of a nonselfadjoint differential operator*, Dokl. Akad. Nauk SSSR **159** (1964), 723–725; English transl. in Soviet Math. Dokl. **5** (1964).

13. A. A. Shkalikov, *On a certain system of functions*, Mat. Zametki **18** (1975), 855–860; English transl. in Math. Notes **18** (1975).

14. G. V. Radzievskiĭ, *On a method of proving completeness of the root vectors of operator-valued functions*, Dokl. Akad. Nauk SSSR **214** (1974), 291–294; English transl. in Soviet Math. Dokl. **15** (1974).

15. Yu. A. Kaz′min, *On the closures of the linear hulls of two systems of functions*, Dokl. Akad. Nauk SSSR **236** (1977), 535–537; English transl. in Soviet Math. Dokl. **18** (1977); also Current Problems of Mathematical Analysis, Izdat Rostov. Gos. Univ., Rostov-on-Don, 1978, pp. 72–83. (Russian)

16. A. G. Tumarkin, *On the completeness of certain systems of functions*, Funktsional. Anal. i Prilozhen. **14** (1980), no. 2, 81–82; English transl. in Functional Anal. Appl. **14** (1980).

17. V. È. Katsnel′son, *A generalization of the Paley–Wiener theorem on the representation of entire functions of finite degree*, Teor. Funktsiĭ, Funktsional. Anal. i Prilozhen. Vyp. **1** (1965), 99–110. (Russian)

Translated by JULIAN GEVIRTZ

Amer. Math. Soc. Transl.
(2) Vol. **144**, 1989

Indefinite Metric in Schur's Interpolation Problem for Analytic Functions. I

UDC 517.54

V. K. DUBOVOĬ

The characteristic operator function (c.o.f.), a fundamental concept introduced by M. S. Livshitz, is a powerful tool for studying contraction operators in Hilbert space. If we define an operator only up to unitary equivalence, the c.o.f. allows us to study contraction operators by analytic methods. The class of c.o.f. of contraction operators admits a simple intrinsic description: it is the class of contraction functions that are analytic in the unit disk.

On the other hand, contraction functions that are analytic in the unit disk are the main topic for study in the Schur interpolation problem. Schur's problem, like many other classical interpolation problems of analysis, attracted the attention of mathematicians from the instant of its formulation. In recent years, the development of the methods of J-theory has allowed V. P. Potapov and a group of his students to make further progress in the study of problems of this kind.

In this connection there has developed an interest in problems that stimulate a synthesis of these two directions. A problem of this type was called to the author's attention by V. D. Potapov. The problem is as follows. It is known that a wide class of interpolation problems can, as follows from a theorem of S. A. Orlov, be classified according to the ranks of the radii of Weyl's limit disks. A similar classification is also possible in Schur's problem, i.e., in the class of c.o.f. of contraction operators. There then arises the natural question of which classification of contraction operators is generated in this context. It turns out that the ranks of the radii of Weyl's limit disks in Schur's problem coincide with the multiplicities of one-sided translations contained in a given contraction operator and its conjugate. It is interesting

1980 *Mathematics Subject Classification* (1985 *Revision*). Primary 30G34, 30E05.
Translation of Teor. Funktsiĭ, Funktsional. Anal. i Prilozhen. Vyp. 37 (1982), 14–26; MR **85f:**30059a.

to notice that the operator situation required more refined methods than the classification of questions in Schur's problem. It turned out that the methods developed for studying c.o.f. widen our possibilities for the study of Schur's problem. Consequently the connection that has been established is useful in both directions.

There are now many articles that develop the methods of J-theory and apply them to the classical interpolation problems of analysis; for example, [1]–[3]. In addition, the Schur problem was solved for square matrix-valued functions in [4] by the methods of J-theory.

However, the consideration of c.o.f. leads in a natural way to the study of Schur's problem for rectangular matrix functions. In this connection, in the present paper we expound Schur's problem for this more general situation, and in the form in which it is needed for discussing the questions mentioned above. Specifically, this goal is pursued in the first part of the present paper. In it we present, mainly without proofs, the known facts for the case of rectangular matrix-valued functions, and along the way we correct some inaccuracies in [4]. In what follows, as we go deeper, problems that were not discussed previously are treated rather thoroughly.

I thank V. P. Potapov, who drew my attention to the present subject, for helpful discussions.

§1. Formulation of the problem. Deduction of the fundamental matrix inequalities

1. Let D be the open unit disk in the complex plane: $D = \{\zeta : |\zeta| < 1\}$. Let $L_{p,q}$ be the set of rectangular matrices, with p rows and q columns, over the complex field. We denote by $S_{p,q}$ the set of matrix-valued functions $\theta(\zeta)$ that are analytic in D with values in $L_{p,q}$ and satisfy the inequality $I_q - \theta^*(\zeta)\theta(\zeta) \geq 0$ for every $\zeta \in D$; in cases when the dimension of the identity matrix is not in doubt, the index q is omitted. Schur's problem [5], generalized to the matrix case, can be formulated as follows.

Given: $n+1$ matrices $c_0, c_1, \dots, c_n$, belonging to $L_{p,q}$.

Required: a) to find a necessary and sufficient condition for $c_0, c_1, \dots, c_n$ to be the leading coefficients in the expansion $\theta(\zeta) = c_0 + c_1\zeta + \dots + c_n\zeta^n + \cdots$ in a series of matrix-valued functions of class $S_{p,q}$; b) to describe all such matrix-valued functions.

2. The approach to Schur's problem that we present here is based on a generalization of Schwarz's lemma to the class $S_{p,q}$. Specifically, we have the following proposition.

THEOREM 1.1. *Let $\zeta_k \in D$, $1 \leq k \leq n$. A necessary and sufficient condition for a matrix-valued function $\theta(\zeta)$ that is analytic in D to belong to $S_{p,q}$ is that* *the inequality*

$$\begin{bmatrix} \dfrac{I-\theta(\zeta_j)\theta^*(\zeta_k)}{1-\zeta_j\overline{\zeta}_k} & \dfrac{I-\theta(\zeta_j)\theta^*(\zeta)}{1-\zeta_j\overline{\zeta}} \\ \dfrac{I-\theta(\zeta)\theta^*(\zeta_k)}{1-\zeta\overline{\zeta}_k} & \dfrac{I-\theta(\zeta)\theta^*(\zeta)}{1-\zeta\overline{\zeta}} \end{bmatrix}_1^n \geq 1 \tag{1.1}$$

is satisfied for $\zeta \in D$.

THEOREM 1.1′. *Let* $\zeta_k \in D$, $1 \leq k \leq n$. *A necessary and sufficient condition for a matrix-valued function that is analytic in D to belong to* $S_{p,q}$ *is that the inequality*

$$\begin{bmatrix} \dfrac{I-\theta(\zeta_j)\theta^*(\zeta_k)}{1-\zeta_j\overline{\zeta}_k} & \dfrac{\theta(\zeta)-\theta(\zeta_j)}{\zeta-\zeta_j} \\ \dfrac{\theta^*(\zeta)-\theta^*(\zeta_k)}{\overline{\zeta}-\overline{\zeta}_k} & \dfrac{I-\theta^*(\zeta)\theta(\zeta)}{1-\overline{\zeta}\zeta} \end{bmatrix}_1^n \geq 0 \tag{1.1′}$$

is satisfied for $\zeta \in D$.

Proofs of (1.1) and (1.1′) are given in [4] for the case $p = q$. To prove these propositions for rectangular matrix functions, it is sufficient to extend the matrix-valued functions to square ones by adjoining zero elements.

Inequality (1.1′) is usually called the dual of (1.1).

3. The next theorem follows from (1.1) in the same way as in [4].

THEOREM 1.2. *A necessary and sufficient condition for a matrix-valued function* $\theta(\zeta) = c_0 + c_1\zeta + \cdots + c_n\zeta^n + \cdots$ *to belong to* $S_{p,q}$ *is that the matrix*

$$C_n = \begin{bmatrix} c_0 & & & 0 \\ c_1 & c_0 & & \\ \vdots & \vdots & \ddots & \\ c_n & c_{n-1} & \cdots & c_0 \end{bmatrix} \tag{1.2}$$

is contractive for every n: $I - C_nC_n^* \geq 0$.

This theorem is completed by the following result.

THEOREM 1.3. *Let* $c_k \in L_{p,q}$, $k = 1, \ldots, n$, *have the property that* $I - C_kC_k^* \geq 0$, *where* C_k *has the form* (1.2). *Then there is a* $\theta(\zeta) \in S_{p,q}$, *the* $n+1$ *leading coefficients of whose series expansion are the matrices* $c_0, \ldots, c_n$: $\theta(\zeta) = c_0 + c_1\zeta + \cdots + c_n\zeta^n + \cdots$.

The proof is based on the following proposition.

If the matrices $c_0, c_1, \ldots, c_n$ satisfy the inequality $A_n = I - C_nC_n^* \geq 0$, then there is a matrix $z \in L_{p,q}$ such that

$$A_{n+1} = I - C_{n+1}C_{n+1}^* \geq 0, \tag{1.3}$$

where

$$C_{n+1} = \begin{bmatrix} c_0 & & & & \\ c_1 & c_0 & & 0 & \\ \vdots & \vdots & \ddots & & \\ c_n & c_{n-1} & \cdots & c_0 & \\ z & c_n & \cdots & c_1 & c_0 \end{bmatrix}.$$

The content of this theorem is that the set of matrices $c_0, \dots, c_n$ can always be extended to a sequence $c_0, c_1, c_2, \dots, c_n, c_{n+1}, c_{n+2}, \dots$ for which the matrices $A_k = I - C_k C_k^* \geq 0$ for all $k = 0, 1, 2, \dots, n, \dots$. But then, by Theorem 1.2, the matrix-valued function $\theta(\zeta) = c_0 + c_1\zeta + \dots + c_n\zeta^n + \cdots$ belongs to $S_{p,q}$.

In [4] there is a mistake in the proof of this proposition. In view of this, we present a proof in more detail.

If we apply the block-matrix lemma [6] to the matrix

$$A_n = I - C_n C_n^* = \begin{bmatrix} A_{n-1} & B_n \\ B_n^* & I - \sum_{k=0}^n c_k c_k^* \end{bmatrix} \geq 0$$

we find that the matrix equation

$$A_{n-1}X = B_n, \qquad X = \begin{bmatrix} x_1 \\ \vdots \\ x_n \end{bmatrix}, \quad x_k \in L_{p,p}, \ 1 \leq k \leq n,$$

has a solution; moreover,

$$d = I - \sum_{k=0}^{n} c_k c_k^* - X^* A_{n-1} X \geq 0.$$

Let $y = c_0[c_1^*, \dots, c_n^*]X$, $z = y^* c_0$. Then if we multiply the matrix (1.3) on the right by

$$T = \begin{bmatrix} I_{(n+1)p} & \begin{bmatrix} -y \\ -X \end{bmatrix} \\ 0 & I_p \end{bmatrix}$$

and on the left by T^*, we obtain, after some simple calculations,

$$T^* A_{n+1} T = \begin{bmatrix} A_n & 0 \\ 0 & d + y^* y \end{bmatrix} \geq 0.$$

Since T is nonsingular, it follows from this that $A_{n+1} \geq 0$. Therefore we need only set

$$c_{n+1} = z = y^* c_0 = X^* \begin{bmatrix} c_1 \\ \vdots \\ c_n \end{bmatrix} c_0^* c_0,$$

where X is the solution of $A_{n-1}X = B_n$.

Theorems 1.2 and 1.3 provide an answer to the first question that was raised in Schur's problem. In fact, the condition $A_n = I - C_n C_n^* \geq 0$, where C_n has the form (1.2), is necessary and sufficient for the matrices $c_0, c_1, \dots, c_n$ to be the leading coefficients in the series $\theta(\zeta) = c_0 + c_1\zeta + \dots + c_n\zeta^n + \cdots$ of matrix functions of class $S_{p,q}$.

4. In our approach, with each classical problem there is associated a fundamental matrix inequality. In this subsection we present the fundamental matrix inequalities connected with Schur's problem.

THEOREM 1.4. *A necessary and sufficient condition for a matrix-valued function $\theta(\zeta)$, analytic in D, to belong to $S_{p,q}$ and have $n+1$ prescribed leading coefficients in the series expansion $\theta(\zeta) = c_0 + c_1\zeta + \cdots + c_n\zeta^n + \cdots$ is that it satisfies the inequality*

$$\left[\begin{array}{c|c} I - C_nC_n^* & \begin{matrix} I - c_0\theta^*(\zeta) \\ \overline{\zeta}(I - (c_0 + c_1/\overline{\zeta})\theta^*(\zeta)) \\ \cdots\cdots\cdots \\ \overline{\zeta}^n(I - (c_0 + c_1/\overline{\zeta} + \cdots + c_n/\overline{\zeta}^n)\theta^*(\zeta)) \end{matrix} \\ \hline * & \frac{I - \theta(\zeta)\theta^*(\zeta)}{1 - \zeta\overline{\zeta}} \end{array}\right] \geq 0 \qquad \text{(S)}$$

for $|\zeta| < 1$.

The necessity of the condition in the theorem can be proved as in [4]. We present the proof of the sufficiency in more detail, since there are errors in the sufficiency proof in [4].

We first call attention to the fact that (S) can be rewritten in the form

$$\left[\begin{array}{c|c} I - C_nC_n^* & \Lambda_{p,n}^*(\zeta) - C_n\Lambda_{q,n}^*(\zeta)\theta^*(\zeta) \\ \hline * & \frac{I - \theta(\zeta)\theta^*(\zeta)}{1 - \overline{\zeta}\zeta} \end{array}\right] \geq 0, \qquad (1.4)$$

where $\Lambda_{p,n}(\zeta) = [I_p, \zeta I_p, \ldots, \zeta^n I_p]$.

If (1.4) is satisfied, for $|\zeta| < 1$ we have

$$\frac{I - \theta(\zeta)\theta^*(\zeta)}{1 - \zeta\overline{\zeta}} \geq 0$$

and consequently $\theta(\zeta)$ belongs to $S_{p,q}$.

Also let $\theta(\zeta) = d_0 + d_1\zeta + d_2\zeta^2 + \cdots$. We need to show that (1.4) implies the equations $d_k = c_k$, $k = 0, 1, 2, \ldots, n$. Let

$$D_\infty = \begin{bmatrix} d_0 & & & & \\ d_1 & d_0 & & 0 & \\ \vdots & \vdots & \ddots & & \\ d_n & d_{n-1} & \ldots & d_0 & \\ d_{n+1} & d_n & \ldots & d_1 & d_0 \\ \vdots & \vdots & & \vdots & \vdots \end{bmatrix}, \qquad D_n = \begin{bmatrix} d_0 & & & \\ d_1 & d_0 & 0 & \\ \vdots & \vdots & \ddots & \\ d_n & d_{n-1} & \ldots & d_0 \end{bmatrix}.$$

Then it is easily seen that

$$\Lambda_q^*(\zeta)\theta^*(\zeta) = D_\infty^*\Lambda_p^*(\zeta), \qquad \frac{I - \theta(\zeta)\theta^*(\zeta)}{1 - \overline{\zeta}\zeta} = \Lambda_p(\zeta)(I - D_\infty D_\infty^*)\Lambda_p^*\zeta,$$

where $\Lambda_p(\zeta) = [I_p, \zeta I_p, \ldots, \zeta^n I_p, \ldots]$. Evidently (1.4) is equivalent to the inequality

$$\left[\begin{array}{c|c} \begin{bmatrix} I - C_nC_n^* & 0 \\ 0 & 0 \end{bmatrix} & \left(\begin{bmatrix} I & 0 \\ 0 & 0 \end{bmatrix} - \begin{bmatrix} C_n & 0 \\ 0 & 0 \end{bmatrix} D_\infty^*\right)\Lambda_p^*(\zeta) \\ \hline * & \Lambda_p(\zeta)(I - D_\infty D_\infty^*)\Lambda_p^*(\zeta) \end{array}\right] \geq 0. \qquad (1.5)$$

If we multiply (1.5) on the left by $T(\zeta) = [\Lambda_p(\zeta), -I_p]$, and on the right by $T^*(\zeta)$, we obtain, after simple transformations, the inequality

$$\Lambda_p(\zeta)(I - QQ^*)\Lambda_p^*(\zeta) \geq (1 + |\zeta|^2 + \cdots + |\zeta|^{2n})I_p, \tag{1.6}$$

where

$$Q = \begin{bmatrix} C_n & 0 \\ 0 & 0 \end{bmatrix} - D_\infty.$$

It follows from (1.6) that

$$\Lambda_p(\zeta)QQ^*\Lambda_p^*(\zeta) \leq \frac{|\zeta|^{2n+2}}{1 - |\zeta|^2} I_p. \tag{1.7}$$

Hence, by using the form of Q, we obtain $c_k = d_k$, $k = 0, 1, 2, \ldots, n$. This is what was to be proved.

THEOREM 1.5. *A necessary and sufficient condition for a matrix function $\theta(\zeta)$, analytic in D, to belong to $S_{p,q}$ and have $n + 1$ prescribed leading coefficients in the series $\theta(\zeta) = c_0 + c_1\zeta + \cdots + c_n\zeta^n + \cdots$ is that the inequality*

$$\left[\begin{array}{c|c} I - C_nC_n^* & \begin{array}{c} \zeta^{-1}(\theta(\zeta) - c_0) \\ \zeta^{-2}(\theta(\zeta) - c_0 - c_1\zeta) \\ \cdots\cdots\cdots \\ \zeta^{-(n+1)}(\theta(\zeta) - c_0 - c_1\zeta - \cdots - c_n\zeta^n) \end{array} \\ \hline * & \frac{I - \theta^*(\zeta)\theta(\zeta)}{1 - \bar{\zeta}\zeta} \end{array}\right] \geq 0 \tag{$\tilde{\mathrm{S}}$}$$

is satisfied for $|\zeta| < 1$.

This proposition is proved exactly as for the case of square matrix-valued functions [4].

The inequalities (S) and ($\tilde{\mathrm{S}}$) are called the fundamental and dual matrix inequalities, respectively, for Schur's problem. By solving either of them we obtain a complete description of all the solutions of Schur's problem.

§2. Elementary multiple factors

1. To describe the set of solutions of the inequalities (S) and ($\tilde{\mathrm{S}}$), we shall need the concept of elementary multiple factors [2], [4].

DEFINITION. An analytic matrix function $B(\zeta)$ of order N which has a single pole in the extended complex plane, of arbitrary multiplicity, is called a *J-elementary multiple factor, J-expansive in the unit disk*, if it is J-expansive outside the disk and J-unitary on its boundary, i.e.,

$$B(\zeta)JB^*(\zeta) - J \geq 0, \qquad |\zeta| < 1, \tag{2.1}$$

$$B(\zeta)JB^*(\zeta) - J = 0, \qquad |\zeta| = 1, \tag{2.2}$$

or, equivalently,

$$B^*(\zeta)JB(\zeta) - J \geq 0, \qquad |\zeta| < 1,$$

$$B^*(\zeta)JB(\zeta) - J = 0, \qquad |\zeta| = 1.$$

Here J is a constant, Hermitian, and involutory matrix of order N: $J^* = J$ and $J^2 = I$.

We shall be interested in the case when the pole of the matrix function is located, depending on the situation, either at $\zeta = 0$ or at $\zeta = \infty$. In addition, in what follows the J-elementary multiple factor will be normalized to unity at $\zeta = 1$, i.e., $B(1) = I$.

With the inequality (S) there is connected a matrix J of the form

$$J = j = \begin{bmatrix} -I_q & 0 \\ 0 & I_p \end{bmatrix}.$$

This is explained by the fact that the matrix element in the lower right-hand corner of the inequality admits a representation in the form

$$I_p - \theta(\zeta)\theta^*(\zeta) = [\theta(\zeta), I_p] \begin{bmatrix} -I_q & 0 \\ 0 & I_p \end{bmatrix} \begin{bmatrix} \theta^*(\zeta) \\ I_p \end{bmatrix}.$$

By the same argument, with inequality (S) there is connected a matrix J of the form

$$J = \tilde{j} = \begin{bmatrix} -I_p & 0 \\ 0 & I_q \end{bmatrix}.$$

2. Let $J = j$ and let j be an elementary multiple factor $B_n(\zeta)$ with a pole at the point O:

$$B_n(\zeta) = d_0 + \frac{d_1}{\zeta} + \cdots + \frac{d_{n+1}}{\zeta^{n+1}}. \tag{2.3}$$

Then it follows from (2.1) that rank $d_{n+1} \le p$. In the case when rank $d_{n+1} = p$, we call $B_n(\zeta)$ a *j-elementary multiple factor of full rank.*

The following theorem is proved as in [4].

THEOREM 2.1 (on parametrization). *For each multiple j-elementary matrix-valued function* (2.3) *of full rank there are uniquely determined parameters $c_0, c_1, \ldots, c_n$, belonging to $L_{p,q}$ and satisfying the condition*

$$A_n = I - C_n C_n^* > 0, \qquad C_n = \begin{bmatrix} c_0 & & & \\ c_1 & c_0 & & 0 \\ \vdots & \vdots & \ddots & \\ c_n & c_{n-1} & \ldots & c_0 \end{bmatrix}$$

such that $B_n(\zeta)$ admits a representation of the form

$$B_n(\zeta) = I + \frac{1-\zeta}{\zeta} j \begin{bmatrix} \Lambda_{q,n}(1) & 0 \\ 0 & \Lambda_{p,n}(1) \end{bmatrix} H_n \begin{bmatrix} \Lambda_{q,n}^*(1/\bar{\zeta}) & 0 \\ 0 & \Lambda_{p,n}^*(1/\bar{\zeta}) \end{bmatrix}, \tag{2.4}$$

$$H_n = \begin{bmatrix} C_n^* \\ I \end{bmatrix} (I - C_n C_n^*)^{-1} [C_n, I]. \tag{2.5}$$

Conversely, every matrix-valued function $B_n(\zeta)$ of this form is a j-elementary multiple function of full rank, and the j-form $B_n(z)$ has the form

$$\frac{1}{1-\zeta\overline{\zeta}}(B_n^*(\zeta)jB_n(\zeta)-j) = \frac{1}{\zeta\overline{\zeta}}\begin{bmatrix}\Lambda_{q,n}(1/\overline{\zeta}) & 0\\ 0 & \Lambda_{p,n}(1/\overline{\zeta})\end{bmatrix} H_n \begin{bmatrix}\Lambda_{q,n}^*(1/\overline{\zeta}) & 0\\ 0 & \Lambda_{p,n}^*(1/\overline{\zeta})\end{bmatrix}. \tag{2.6}$$

COROLLARY. *For every binomial j-elementary matrix-valued function of full rank, $b_0(\zeta) = I + ((1-\zeta)/\zeta)P_0$, with pole at $\zeta = 0$, the parameter $c_0 \in L_{p,q}$ is uniquely determined so that $I - c_0c_0^* > 0$ and $b_0(\zeta)$ admits the representation*

$$b_0(\zeta) = I + \frac{1-\zeta}{\zeta}j\begin{bmatrix}c_0^*\\ I\end{bmatrix}(I - c_0c_0^*)^{-1}[c_0, I]. \tag{2.7}$$

Conversely, every matrix-valued function of this form is a j-elementary function of full rank.

Consequently

$$P_0 = \begin{bmatrix}\xi_0^*\\ \eta_0^*\end{bmatrix}[\xi_0, \eta_0]j \tag{2.8}$$

where

$$\xi_0 = -(I - c_0c_0^*)^{-1/2}c_0, \qquad \eta_0 = (I - c_0c_0^*)^{-1/2}. \tag{2.9}$$

Here we note that P_0 has the following properties:

$$P_0^2 = P_0, \qquad P_0j \geq 0, \qquad \operatorname{rank} P_0 = p. \tag{2.10}$$

3. For the solution of $(\tilde{\mathrm{S}})$ we need a j-elementary multiple factor of full rank with pole at ∞:

$$\tilde{B}_n(\zeta) = \tilde{d}_0 + \tilde{d}_1\zeta + \cdots + \tilde{d}_{n+1}\zeta^{n+1}, \qquad \operatorname{rank} \tilde{d}_{n+1} = p. \tag{2.11}$$

Evidently $\tilde{B}_n^*(1/\overline{\zeta})$ is a $\tilde{j}$-elementary multiple factor of full rank with pole at $\zeta = 0$. Therefore it follows from Theorem 2.1 on parametrization that the matrices $c_k \in L_{p,q}$, $0 \leq k \leq n$, are uniquely determined so that

$$\tilde{B}_n^*\left(\frac{1}{\overline{\zeta}}\right) = I - \frac{1-\zeta}{\zeta}\tilde{j}\begin{bmatrix}\Lambda_{p,n}(1) & 0\\ 0 & \Lambda_{q,n}(1)\end{bmatrix}\tilde{H}_n\begin{bmatrix}\Lambda_{p,n}^*(1/\overline{\zeta}) & 0\\ 0 & \Lambda_{q,n}^*(1/\overline{\zeta})\end{bmatrix}, \tag{2.12}$$

where

$$\tilde{H}_n = \begin{bmatrix}I\\ C_n^*\end{bmatrix}(I - C_nC_n^*)^{-1}[I, C_n]. \tag{2.13}$$

Hence we obtain

$$\tilde{B}_n(\zeta) = I + (1-\zeta)\begin{bmatrix}\Lambda_{p,n}(\zeta) & 0\\ 0 & \Lambda_{q,n}(\zeta)\end{bmatrix}\tilde{H}_n\begin{bmatrix}\Lambda_{p,n}^*(1) & 0\\ 0 & \Lambda_{q,n}^*(1)\end{bmatrix}\tilde{j}. \tag{2.14}$$

Consequently we have established the first part of the following proposition.

THEOREM 2.2. *For every $\tilde{j}$-elementary multiple matrix-valued function* (2.11) *of full rank there are uniquely determined parameters $c_k \in L_{p,q}$, $0 \le k \le n$, that satisfy the condition $A_n = I - C_n C_n^* > 0$ and are such that $\tilde{B}_n(\zeta)$ admits a representation of the form* (2.14).

Conversely, every matrix-valued function of the form (2.14) *is a j-elementary multiple matrix-valued function of full rank, and the $\tilde{j}$-form $\tilde{B}_n(\zeta)$ has the form*

$$\frac{1}{1-\zeta\overline{\zeta}}(\tilde{B}_n(\zeta)\tilde{j}\tilde{B}_n^*(\zeta)-\tilde{j}) = \begin{bmatrix} \Lambda_{p,n}(\zeta) & 0 \\ 0 & \Lambda_{q,n}(\zeta) \end{bmatrix} \tilde{H}_n \begin{bmatrix} \Lambda_{p,n}^*(\zeta) & 0 \\ 0 & \Lambda_{q,n}^*(\zeta) \end{bmatrix}. \tag{2.15}$$

The second part of this proposition can be proved in the same way as for Theorem 2.1.

COROLLARY. *For every binomial $\tilde{j}$-elementary matrix-valued function $\tilde{b}_0(\zeta) = I + (1-\zeta)Q_0$ of full rank, there is a uniquely determined matrix $c_0 \in L_{p,q}$ such that $I - c_0 c_0^* > 0$ and $\tilde{b}_0(\zeta)$ admits the representation*

$$\tilde{b}_0(\zeta) = I + (1-\zeta)\begin{bmatrix} I \\ c_0^* \end{bmatrix}(I - c_0c_0^*)^{-1}[I, c_0]\tilde{j}. \tag{2.16}$$

4. Let $B_k(\zeta)$, $0 \le k \le n$, be the j-elementary multiple factors corresponding to the parameters $c_0, c_1, \dots, c_k$, $0 \le k \le n$. Applying the reasoning which has become familiar in the present approach, we can show that $B_{k+1}(\zeta)$ is divisible by $B_k(\zeta)$, $0 \le k \le n$, and moreover that the quotient is a j-elementary binomial factor of full rank. This means that $B_n(\zeta)$ can be decomposed into a product of j-elementary binomial factors of full rank:

$$B_n(\zeta) = b_n(\zeta)b_{n-1}(\zeta)\cdots b_1(\zeta)b_0(\zeta). \tag{2.17}$$

The resulting binomial factors $b_k(\zeta)$, $0 \le k \le n$, can be expressed directly in terms of the parameters $c_0, c_1, \dots, c_n$:

$$\begin{aligned} b_k(\zeta) = I + \frac{1-\zeta}{\zeta} j \Bigg(&\begin{bmatrix} \Lambda_{q,k}(0) & 0 \\ 0 & \Lambda_{p,k}(1) \end{bmatrix} H_k \begin{bmatrix} \Lambda_{q,k}^*(1) & 0 \\ 0 & \Lambda_{p,k}^*(1) \end{bmatrix} \\ &- \begin{bmatrix} \Lambda_{q,k-1}(1) & 0 \\ 0 & \Lambda_{p,k-1}(1) \end{bmatrix} H_{k-1} \begin{bmatrix} \Lambda_{q,k-1}^*(1) & 0 \\ 0 & \Lambda_{p,k-1}^*(1) \end{bmatrix} \Bigg), \end{aligned} \tag{2.18}$$

and $b_0(\zeta)$ has the form (2.7).

In a similar way we can obtain the decomposition

$$\tilde{B}_n(\zeta) = \tilde{b}_0(\zeta)\tilde{b}_1(\zeta)\cdots\tilde{b}_n(\zeta),$$

where, for $k = 1, 2, \dots, n$,

$$\begin{aligned} \tilde{b}_k(\zeta) = I + (1-\zeta) \Bigg(&\begin{bmatrix} \Lambda_{p,k}(1) & 0 \\ 0 & \Lambda_{q,k}(1) \end{bmatrix} \tilde{H}_k \begin{bmatrix} \Lambda_{p,k}^*(1) & 0 \\ 0 & \Lambda_{q,k}^*(1) \end{bmatrix} \\ &- \begin{bmatrix} \Lambda_{p,k-1}(1) & 0 \\ 0 & \Lambda_{q,k-1}(1) \end{bmatrix} \tilde{H}_{k-1} \begin{bmatrix} \Lambda_{p,k-1}^*(1) & 0 \\ 0 & \Lambda_{q,k-1}^*(1) \end{bmatrix} \Bigg) \tilde{j} \end{aligned} \tag{2.19}$$

and $\tilde{b}_0(\zeta)$ has the form (2.16).

§3. Solution of the fundamental matrix inequalities

We shall solve the fundamental matrix inequalities (S) and ($\tilde{\text{S}}$) under the assumption of the nondegeneracy of the so-called fundamental information block $A_n = I - C_n C_n^* > 0$. We shall say that such a Schur problem is *nondegenerate.* The case of a degenerate information block requires special consideration, and will be discussed in a separate paper. A full description of the set of solutions of the fundamental matrix equations in the nondegenerate case is given by the following two propositions.

THEOREM 3.1. *The general solution $\theta(\zeta)$ of the fundamental matrix inequality* (S) *is representable as a linear fractional transformation of an arbitrary matrix function $\omega(\zeta) \in S_{p,q}$:*

$$\theta(\zeta) = [\omega(\zeta)b(\zeta) + d(\zeta)]^{-1}[\omega(\zeta)a(\zeta) + c(\zeta)] \tag{3.1}$$

whose coefficient matrix is constructed from the matrices C_n:

$$\begin{aligned} B_n(\zeta) &= \begin{bmatrix} a(\zeta) & b(\zeta) \\ c(\zeta) & d(\zeta) \end{bmatrix} \\ &= I + \frac{1-\zeta}{\zeta} j \begin{bmatrix} \Lambda_{q,n}(1) & 0 \\ 0 & \Lambda_{p,n}(1) \end{bmatrix} H_n \begin{bmatrix} \Lambda_{q,n}^*(1/\overline{\zeta}) & 0 \\ 0 & \Lambda_{p,n}^*(1/\overline{\zeta}) \end{bmatrix} \end{aligned} \tag{3.2}$$

and is a j-expansive elementary multiple factor of full rank with a pole of multiplicity $n+1$ at the point $\zeta = 0$.

THEOREM 3.2. *The general solution $\theta(\zeta)$ of the fundamental matrix inequality* ($\tilde{\text{S}}$) *is representable as a linear fractional transformation of an arbitrary matrix function $\omega(\zeta) \in S_{p,q}$:*

$$\theta(\zeta) = [\tilde{a} \times (\zeta)\omega(\zeta) + \tilde{b}(\zeta)][\tilde{c}(\zeta)\omega(\zeta) + \tilde{d}(\zeta)]^{-1} \tag{3.3}$$

whose coefficient matrix is constructed from the matrices C_n:

$$\begin{aligned} \tilde{B}_n(\zeta) &= \begin{bmatrix} \tilde{a}(\zeta) & \tilde{b}(\zeta) \\ \tilde{c}(\zeta) & \tilde{d}(\zeta) \end{bmatrix} \\ &= I + (1-\zeta) \begin{bmatrix} \Lambda_{p,n}(\zeta) & 0 \\ 0 & \Lambda_{q,n}(\zeta) \end{bmatrix} \tilde{H}_n \begin{bmatrix} \Lambda_{p,n}^*(1) & 0 \\ 0 & \Lambda_{q,n}^*(1) \end{bmatrix} \tilde{j} \end{aligned} \tag{3.4}$$

and is a $\tilde{j}$-expansive elementary multiple factor of full rank with a pole of multiplicity $n+1$ at the point $\zeta = \infty$.

These propositions are proved exactly as for square matrices [4]. Notice that the main step is the proof that the inequalities (S) and ($\tilde{\text{S}}$) can be rewritten in the respective forms

$$[\theta(\zeta), I] \frac{B_n^{-1}(\zeta) j B_n^{*-1}(\zeta)}{1 - |\zeta|^2} \begin{bmatrix} \theta^*(\varepsilon) \\ I \end{bmatrix} \geq 0, \tag{3.5}$$

$$[\theta^*(\zeta), I] \frac{\tilde{B}_n^{*-1}(\zeta) \tilde{j} \tilde{B}_n^{-1}(\zeta)}{1 - |\zeta|^2} \begin{bmatrix} \theta(\zeta) \\ I \end{bmatrix} \geq 0. \tag{3.6}$$

Then, decomposing the matrix-valued functions $B_n(\zeta)$ and $\tilde{B}_n(\zeta)$ into blocks, we obtain the solutions of these inequalities in the forms (3.1) and (3.3).

Let

$$\tilde{B}_n(\zeta)\{\omega(\zeta)\} = [\tilde{a}(\zeta)\omega(\zeta) + \tilde{b}(\zeta)][\tilde{c}(\zeta)\omega(\zeta) + \tilde{d}(\zeta)]^{-1};$$

then (3.3) can be rewritten in the form $\theta(\zeta) = \tilde{B}_n(\zeta)\{\omega(\zeta)\}$. Here, as above, any function $\omega(\zeta) \in S_{p,q}$ can serve as the parameter.

Consequently all the information about the Schur problem is contained in the multiple factor $\tilde{B}_n(\zeta)$ of full rank, and every property of this function is reflected in the problem. That is, the nondegeneracy of the Schur problem is equally determined by the factor $\tilde{B}_n(\zeta)$. Evidently a similar inference can be made for the multiple factor $B_n(\zeta)$.

§4. Stepwise solution of Schur's problem. Schur parameters

Here, following the remarkable idea of Schur [5], we solve the Schur problem recursively, supposing as before that

$$A_n = I - C_n C_n^* > 0. \tag{4.1}$$

1. We first solve the problem in the simplest case, when only the constant term c_0 is given. It follows from Theorem 3.2 that the set of matrix-valued functions $\theta(\zeta) \in S_{p,q}$ have the matrix c_0 as constant term in the power series expansion $\theta(\zeta) = c_0 + \cdots$ determined by the equation

$$\theta(\zeta) = \tilde{b}_0(\zeta)\{\omega(\zeta)\} \tag{4.2}$$

where $\tilde{b}_0(\zeta)$ has the form (2.13) and $\omega(\zeta) \in S_{p,q}$.

2. Now we look among all the solutions $\theta(\zeta)$ so obtained for those that have two given coefficients in the series $\theta(\zeta) = c_0 + c_1\zeta + \cdots$. Each such function $\theta(\zeta)$ admits a representation of the form (4.2). Let us describe the set of parameters $\omega(\zeta)$ obtained in this way. It is convenient to set $\theta_1(\zeta) = \omega(\zeta)$ and let $\theta_1(\zeta) = c_0^{(1)} + \cdots$. Then if we rewrite (4.2) in the form

$$\theta(\zeta) = \tilde{b}_0(\zeta)\{\theta_1(\zeta)\}, \tag{4.3}$$

it follows that

$$[I + c_1 c_0^*(I - c_0 c_0^*)^{-1}]c_0^{(1)} = [I + c_1 c_0^*(I - c_0 c_0^*)^{-1}]c_0 + c_1.$$

We find from (4.1) that the matrix $I + c_1 c_0^*(I - c_0 c_0^*)^{-1}$ has an inverse. Therefore

$$c_0^{(1)} = c_0 + [I + c_1 c_0^*(I - c_0 c_0^*)^{-1}]^{-1} c_1. \tag{4.4}$$

Hence we obtain

$$I - c_0^{(1)} c_0^{(1)*} > 0. \tag{4.5}$$

Consequently the condition $\theta_1(\zeta) = c_0^{(1)} + \cdots$ is necessary for a function $\theta_1(\zeta) \in S_{p,q}$, defined by (4.3), to be a solution $\theta(\zeta)$ of the Schur problem with two given coefficients in the series $\theta(\zeta) = c_0 + c_1\zeta + \cdots$ with $A_1 = I - CC_1^* > 0$; here $c_0^{(1)}$ is determined by (4.4), and (4.5) is satisfied.

It can be shown by simple calculations that the preceding conditions are not only necessary, but also sufficient.

If we define, in the preceding way, all the functions $\theta_1(\zeta) \in S_{p,q}$ of the form $\theta_1(\zeta) = c_0^{(1)} + \cdots$, with $I - c_0^{(1)}c_0^{(1)*} > 0$, we obtain a representation of $\theta_1(\zeta)$ in the form of a linear fractional transformation of an arbitrary function $\theta_2(\zeta) \in S_{p,q}$: $\theta_1(\zeta) = \tilde{b}_1(\zeta)\{\theta_2(\zeta)\}$, where $\tilde{b}_1(\zeta)$ has the form (2.13) with c_0 replaced by $c_0^{(1)}$.

The superposition $\theta(\zeta) = \tilde{b}_0(\zeta)\{\tilde{b}_1(\zeta)\{\theta_2(\zeta)\}\}$, $\theta_2(\zeta) \in S_{p,q}$, with coefficient matrix $\tilde{b}_0(\zeta)\tilde{b}_1(\zeta)$, is the general form of a matrix-valued function $\theta(\zeta)$ with two given coefficients in the series $\theta(\zeta) = c_0 + c_1\zeta + \cdots$.

3. Evidently we can continue this procedure. After $n+1$ steps we obtain the following result.

The general form of a matrix-valued function $\theta(\zeta) \in S_{p,q}$ with $n+1$ prescribed leading coefficients in its power series $\theta(\zeta) = c_0 + c_1\zeta + \cdots + c_n\zeta^n + \cdots$, with $A_n = I - C_nC_n^* > 0$, is a superposition

$$\theta(\zeta) = \tilde{b}_0(\zeta)\{\tilde{b}_1(\zeta)\{\cdots\{\tilde{b}_n(\zeta)\{\theta_{n+1}(\zeta)\}\}\cdots\}\},$$

with an arbitrary $\theta_{n+1}(\zeta) \in S_{p,q}$ as parameter, where

$$\tilde{b}_k(\zeta) = I + (1-\zeta)\begin{bmatrix} I \\ c_0^{(k)*} \end{bmatrix}(I - c_0^{(k)}c_0^{(k)*})^{-1}[I, c_0^{(k)}]\tilde{j}, \tag{4.6}$$

and $c_0^{(0)} = c_0$. The coefficient matrix of the resulting linear fractional transformation has the form

$$\prod_{k=0}^{\overset{n}{\curvearrowright}} \tilde{b}_k(\zeta). \tag{4.7}$$

Consequently, to the Schur problem, under the condition $A_n > 0$, there corresponds the product (4.7) of binomial factors (4.6) of full rank.

The converse is also valid: to an arbitrary product (4.7) of binomial factors (4.6) of full rank there completely corresponds a definite Schur problem.

Thus a Schur problem with a strictly positive information block (4.1) is the same as that of finding a finite product (4.7) of binomial $\tilde{j}$-elementary factors of full rank with poles at the point $\zeta = \infty$.

We note that the parameters $c_0^{(0)}, c_0^{(1)}, \ldots, c_0^{(n)}$ are called the *Schur parameters.*

4. We turn to the connection between the product (4.7) and the multiple factor $\tilde{B}_n(\zeta)$ corresponding to the Schur problem. The product (4.7) is a $\tilde{j}$-elementary multiple matrix-valued function of full rank. By the parametrization theorem such a matrix-valued function can be represented in the form (2.11):

$$\prod_{k=0}^{\overset{n}{\curvearrowright}} \tilde{b}_k(\zeta) = I + (1-\zeta)\begin{bmatrix} \Lambda_{p,n}(\zeta) & 0 \\ 0 & \Lambda_{q,n}(\zeta) \end{bmatrix}\tilde{H}_n\begin{bmatrix} \Lambda_{p,n}^*(1) & 0 \\ 0 & \Lambda_{q,n}^*(1) \end{bmatrix}\tilde{j},$$

where the matrix $\tilde{H}_n$ is determined by the matrices $c_0, c_1, \ldots, c_n$ that coincide with the data of the problem. Therefore

$$\overset{\curvearrowright}{\prod_{k=0}^{n}} \tilde{b}_k(\zeta) = \tilde{B}_n(\zeta).$$

This connection makes it possible, in particular, to express, by means of the formulas (2.16), the binomial factors $\tilde{b}_k(\zeta)$ which are constructed in terms of the Schur parameters, directly in terms of the data of the problem.

5. We have been starting from the inequality ($\tilde{\mathrm{S}}$). If we start from (S) we obtain the same results, but in terms of j-elementary factors of the form (2.7). In particular, the Schur problem with a nondegenerate block (see (4.1)) is the same as that of finding a finite product $\overset{\curvearrowleft}{\prod}_0^n b_k(\zeta)$ of binomial j-elementary factors of full rank with poles at $\zeta = 0$:

$$b_k(\zeta) = I + \frac{1-\zeta_j}{\zeta} \begin{bmatrix} c_0^{(k)*} \\ I \end{bmatrix} (I - c_0^{(k)} c_0^{(k)*})^{-1} [c_0^{(k)}, I], \quad k = 0, 1, \ldots, n, \tag{4.8}$$

where $c_0^{(0)}, c_0^{(1)}, \ldots, c_0^{(n)}$ are the same Schur parameters.

6. The results that we have obtained let us place in correspondence, on the one hand, every matrix-valued function $\theta(\zeta) \in S_{p,q}$, where $\theta(\zeta) = c_0 + c_1\zeta + \cdots + c_n\zeta^n + \cdots$, for which all the information blocks are strictly positive: $A_n > 0$, $n = 0, 1, 2, \ldots$, with, on the other hand, an infinite product of binomial factors (4.6), (4.8) of full rank:

$$\overset{\curvearrowright}{\prod_{k=0}^{\infty}} \tilde{b}_k(\zeta) \left(\overset{\curvearrowleft}{\prod_{k=0}^{\infty}} b_k(\zeta) \right). \tag{4.9}$$

Conversely, to every infinite product of factors of the form (4.6), (4.8), there corresponds a completely determined matrix-valued function $\theta(\zeta) \in S_{p,q}$.

In the Schur problem, the Blaschke–Potapov products always diverge. A precise description of their divergence is presented in terms of the radii of the Weyl disks, which are discussed in the next part of this paper.

References

1. I. V. Kovalishina and V. P. Potapov, *Indefinite metric in the Nevanlinna-Pick problem*, Akad. Nauk Armyan. SSR Dokl. **59** (1974), 17–22; English transl. in Amer. Math. Soc. Transl. (2) **138** (1988).

2. I. V. Kovalishina, *J-expansive matrix-valued functions in the Carathéodory problem*, Akad. Nauk Armyan. SSR Dokl. **59** (1974), 129–135. (Russian)

3. ——, *J-expansive matrix-valued functions and the classical moment problem*, Akad. Nauk Armyan. SSR Dokl. **60** (1975), 3–10. (Russian)

4. L. A. Galstyan, *Analytic j-expanding matrix-valued functions and the Schur problem*, Izv. Akad. Nauk Armyan. SSR. Ser. Mat. **12** (1977), 204–228. (Russian)

5. J. Schur [Issai Schur], *Über Potenzreihen, die im Innern der Einheitskreises beschränkt sind.* I, II, J. Reine Angew. Math. **147** (1917), 205–232; **148** (1918), 122–145.

6. A. V. Efimov and V. P. Potapov, *J-expansive matrix-valued functions and their role in the analytical theory of electric circuits*, Uspekhi Mat. Nauk **28** (1973), no. 1(169), 65–130; English transl. in Russian Math. Surveys **28** (1973).

Translated by RALPH P. BOAS

Amer. Math. Soc. Transl.
(2) Vol. **144**, 1989

Indefinite Metric in Schur's Interpolation Problem for Analytic Functions. II

UDC 519.210

V. K. DUBOVOĬ

§5. Weyl disks in the Schur problem

We now present a geometric interpretation of the set of solutions of Schur's problem.

As we noticed in §3 (see (3.5), (3.6)), the set of solutions of Schur's problem is characterized by the inequality

$$[\theta(\zeta), I]B_n^{-1}(\zeta)jB_n^{*-1}(\zeta)\begin{bmatrix}\theta^*(\zeta)\\ I\end{bmatrix} \geq 0 \tag{5.1}$$

or

$$[\theta^*(\zeta), I]\tilde{B}_n^{*-1}(\zeta)\tilde{j}\tilde{B}_n^{-1}(\zeta)\begin{bmatrix}\theta(\zeta)\\ I\end{bmatrix} \geq 0 \tag{5.2}$$

where $B_n(\zeta)$ and $\tilde{B}_n(\zeta)$ are defined respectively by (3.2) and (3.4). We introduce the notation

$$W_n = B_n^{-1}(\zeta)jB_n^{*-1}(\zeta), \tag{5.3}$$

$$\tilde{W}_n = \tilde{B}_n^{*-1}(\zeta)\tilde{j}\tilde{B}_n^{-1}(\zeta) \tag{5.4}$$

and will call the matrices W_n and $\tilde{W}_n$ the *Weyl matrices*, following [2]. As in (3.2) and (3.4) we decompose the Weyl matrices into blocks:

$$W_n = \begin{pmatrix}-R_n & S_n^*\\ S_n & -T_n\end{pmatrix}, \qquad \tilde{W}_n = \begin{pmatrix}-\tilde{R}_n & \tilde{S}_n\\ \tilde{S}_n^* & -\tilde{T}_n\end{pmatrix}. \tag{5.5}$$

Taking account of (3.2) and (3.4), we find from (5.3) and (5.4) that

$$R_n = I_q + (1 - |\zeta|^2)\Lambda_{q,n}(\zeta)C_n^*A_n^{-1}C_n\Lambda_{q,n}^*(\zeta); \tag{5.6}$$

$$\tilde{R}_n = I_p + \frac{1-|\zeta|^2}{|\zeta|^2}\Lambda_{p,n}\left(\frac{1}{\bar{\zeta}}\right)A_n^{-1}\Lambda_{p,n}^*\left(\frac{1}{\bar{\zeta}}\right), \tag{5.7}$$

1980 *Mathematics Subject Classification* (1985 *Revision*). Primary 30G34, 30E05.

Translation of Teor. Funktsiĭ, Funktsional. Anal. i Prilozhen. Vyp. 38 (1982), 32–39; MR **85f**:30059. Part I is the immediately preceding paper in this volume.

where, as before, $A_n = (I - C_n C_n^*)^{-1}$. It follows from (5.6) and (5.7) that the matrices R_n and $\tilde{R}_n$ are strictly positive:

$$R_n \geq I, \qquad \tilde{R}_n > I, \tag{5.8}$$

at each given point ζ_0 of the open unit disk (different from zero for $\tilde{R}_n$).

It follows from the forms (3.2) and (3.4) of the matrix-valued functions $B_n(\zeta)$ and $\tilde{B}_n(\zeta)$ that

$$\tilde{B}_n(\zeta) = J^{-1} j B_n^{-1}(\zeta) j J, \qquad J = \begin{pmatrix} 0 & I_q \\ I_p & 0 \end{pmatrix}.$$

Hence we obtain $\tilde{W}_n^{-1} = -J^{-1} j W_n j J$.

Passing to the block form (5.5), we obtain

$$-\begin{pmatrix} T_n & S_n \\ S_n^* & R_n \end{pmatrix} \begin{pmatrix} \tilde{R}_n & -\tilde{S}_n \\ -\tilde{S}_n^* & \tilde{T}_n \end{pmatrix} = \begin{pmatrix} I_p & 0 \\ 0 & I_q \end{pmatrix}. \tag{5.9}$$

Hence, using (5.8), we obtain the equations

$$R_n^{-1} = \tilde{S}_n^* \tilde{R}_n^{-1} \tilde{S}_n - \tilde{T}_n, \tag{5.10}$$
$$\tilde{R}_n^{-1} = S_n R_n^{-1} S_n^* - T_n, \tag{5.11}$$
$$\tilde{R}_n^{-1} \tilde{S}_n = S_n R_n^{-1}. \tag{5.12}$$

In the present case, all conditions that are necessary for the construction of the Weyl theory [2] are satisfied, specifically:

1) every matrix W_n is representable in the form

$$W_n = B_n^{-1}(\zeta) j B_n^{*-1}(\zeta) = \begin{pmatrix} -R_n & S_n^* \\ S_n & -T_n \end{pmatrix},$$

where $B_n(\zeta)$ is a nonsingular j-expansive matrix function (for $|\zeta| < 1$);

2) the j-forms $j - B_n^{-1}(\zeta) j B_n^{*-1}(\zeta)$ are monotone nondecreasing as n increases;

3) the block R_n is nonsingular.

Therefore, as in [2], we can prove the following proposition:

THEOREM 5.1. *The set of matrices $\theta \in L_{p,q}$ that satisfy the inequality*

$$[\theta, I] W_n \begin{bmatrix} \theta^* \\ I \end{bmatrix} \geq 0 \tag{5.13}$$

form a matrix disk K_n: $\theta = C_n + \rho_{g,n}^{1/2} u \rho_{d,n}^{1/2}$, $uu^* \leq I$, $u \in L_{p,q}$ with center $C_n = S_n R_n^{-1}$, right radius $\rho_{d,n} = R_n^{-1}$, and left radius $\rho_{g,n} = S_n R_n^{-1} S_n^* - T_n$.

THEOREM 5.2. *As n increases, the disks K_n are nested, with both the right and left radii decreasing monotonically.*

Here the Weyl disk (5.13) is described in terms of the matrices W_n. A similar description can be given in terms of $\tilde{W}_n$.

These results allow us to make a correspondence between matrix functions $\theta(\zeta) \in S_{p,q}$, with $\theta(\zeta) = c_0 + c_1\zeta + \cdots + c_n\zeta^n + \cdots$, for which $A_n > 0$,

$n = 0, 1, \ldots$, and, at each point $\zeta \in D$, a sequence of nested Weyl disks $K_0 \subset K_1 \subset \cdots \subset K_n \subset \cdots$. Letting $n \to \infty$, we obtain the Weyl limit disk

$$\theta = C_\infty + \rho_{g,\infty}^{1/2} u \rho_{d,\infty}^{1/2}, \qquad uu^* \leq I, \ u \in L_{p,q}, \tag{5.14}$$

where

$$C_\infty = \lim_{n\to\infty} C_n, \qquad \rho_{g,\infty} = \lim_{n\to\infty} \rho_{g,n}, \qquad \rho_{d,\infty} = \lim_{n\to\infty} \rho_{d,n}.$$

Notice that in the present case

$$\rho_{g,\infty} = 0. \tag{5.15}$$

In fact, by Theorem 5.1, and equations (5.7) and (5.11), we obtain

$$\rho_{g,n}^{-1}(\zeta) = I_p + \frac{1-|\zeta|^2}{|\zeta|^2} \Lambda_{p,n}\left(\frac{1}{\overline{\zeta}}\right) A_n^{-1} \Lambda_{p,n}^*\left(\frac{1}{\overline{\zeta}}\right). \tag{5.16}$$

Since $A_n^{-1} \geq 1$, we have

$$\rho_{g,n}^{-1}(\zeta) \geq I_p + \frac{1-|\zeta|^2}{|\zeta|^2} \Lambda_{p,n}\left(\frac{1}{\overline{\zeta}}\right) \Lambda_{p,n}^*\left(\frac{1}{\overline{\zeta}}\right) = \frac{1}{|\zeta|^{2n+2}} I_p.$$

Then (5.15) follows from this. This fully agrees with the fact that the "infinite" Schur problem has a unique solution.

The limit Weyl disk K_∞ is shrunk to a point since its left radius has become zero. However, the specific nature of the Schur problem, considered in the matrix case, is distinctive in that although the left radius $\rho_{g,\infty} = 0$, the Weyl limit disk K_∞ is characterized by its right radius $\rho_{d,\infty}$ which, as subsequent discussion shows, is different from zero.

§6. Study of the rank of the radii of the Weyl limit disk

1. From Theorem 5.1 and (5.6) we have

$$\rho_{d,n}^{-1}(\zeta) = I_q + (1-|\zeta|^2)\Lambda_{q,n}(\zeta) C_n^* A_n^{-1} C_n \Lambda_{q,n}^*(\zeta). \tag{6.1}$$

Hence, recalling that $C_n^* A_n^{-1} C_n = (I - C_n^* C_n) - I$, we obtain

$$\rho_{d,n}^{-1}(\zeta) = |\zeta|^{2n+2} I_q + (1-|\zeta|^2)\Lambda_{q,n}(\zeta)(I - C_n^* C_n)^{-1} \Lambda_{q,n}^*(\zeta). \tag{6.2}$$

The subsequent discussion is based on establishing a connection between the radii of the Weyl disks for the functions $\theta(\zeta) \in S_{p,q}$ and $\hat{\theta}(\zeta) = \theta^*(\overline{\zeta}) \in S_{q,p}$. Hence, when necessary, the radii of the Weyl disks for $\theta(\zeta)$ will be denoted by $\rho_{g,n}(\zeta, \theta)$ and $\rho_{d,n}(\zeta, \theta)$.

DEFINITION. The matrix function

$$r_{g,n}(\zeta, \theta) = (1/|\zeta|^{2n+2})\rho_{g,n}(\zeta, \theta), \quad 0 < |\zeta| < 1, \tag{6.3}$$

will be called the *normalized left radius*.

This definition is justified by the following result.

THEOREM 6.1. *We have the equations*

$$\rho_{d,n}(\zeta,\hat{\theta}) = r_{g,n}(\overline{\zeta},\theta), \qquad n = 0,1,2,\dots. \tag{6.4}$$

PROOF. If $\theta(\zeta) = c_0 + c_1\zeta + \cdots + c_n\zeta^n + \cdots$, then $\hat{\theta}(\zeta) = c_0^* + c_1^*\zeta + \cdots + c_n^*\zeta^n + \cdots$. Therefore the information blocks for $\hat{\theta}(\zeta)$ have the form $A_n = I - \hat{C}_n\hat{C}_n^*$, where

$$\hat{C}_n = \begin{pmatrix} c_0^* & & & \\ c_1^* & c_0^* & & 0 \\ \vdots & \vdots & \ddots & \\ c_n^* & c_{n-1}^* & \cdots & c_0^* \end{pmatrix}.$$

If $\theta(\zeta) \in S_{p,q}$ then evidently

$$C_n = U_q C_n^* U_p, \qquad U_p = \begin{pmatrix} 0 & & I_p \\ & \cdot^{\cdot^{\cdot}} & \\ I_p & & 0 \end{pmatrix}. \tag{6.5}$$

From (6.2), by using (6.5), we obtain

$$\rho_{d,n}^{-1}(\zeta,\hat{\theta}) = |\zeta|^{2n+2}I_p + (1-|\zeta|^2)\Lambda_{p,n}(\zeta)U_p(I - C_nC_n^*)^{-1}U_p\Lambda_{p,n}^*(\zeta).$$

Since

$$(1/\zeta^n)\Lambda_{p,n}(\zeta)U_p = \Lambda_{p,n}(1/\zeta),$$

we have

$$(1/|\zeta|^{2n+2})\rho_{d,n}^{-1}(\zeta,\hat{\theta}) = I_p + ((1-|\zeta|)^2/|\zeta|^2)\Lambda_{p,n}(1/\zeta)A_n^{-1}\Lambda_{p,n}^*(1/\zeta).$$

Comparing this with (5.16), we obtain

$$\frac{1}{|\zeta|^{2n+2}}\rho_{d,n}^{-1}(\zeta,\hat{\theta}) = \rho_{g,n}^{-1}(\overline{\zeta},\theta).$$

If we now use (6.3) we arrive at (6.4).

Equation (6.4) lets us define $r_{g,n}(\zeta,\theta)$ at $\zeta = 0$ by setting $r_{g,n}(0,\theta) = \rho_{d,n}(0,\hat{\theta})$. Moreover, it follows from the theorem that we have proved that the limit $r_{g,\infty}(\zeta,\theta) = \lim_{n\to\infty} r_{g,n}(\zeta,\theta)$, $|\zeta| < 1$, exists, and we have the equation

$$\rho_{d,\infty}(\zeta,\hat{\theta}) = r_{g,\infty}(\overline{\zeta},\theta), \qquad |\zeta| < 1. \tag{6.6}$$

We call the matrix function $r_{g,\infty}(\zeta,\theta)$ the *normalized left radius of the Weyl limit disk.*

THEOREM 6.2. *We have the equations*

$$\det \rho_{d,n}(\zeta,\theta) = \det r_{g,n}(\zeta,\theta), \qquad n = 0,1,2,\dots. \tag{6.7}$$

PROOF. It follows from (5.5) that the Weyl matrix W_n can be reduced to diagonal form by means of a triangular transformation:

$$W_n = \begin{bmatrix} I_q & 0 \\ -S_nR_n^{-1} & I_p \end{bmatrix} \begin{bmatrix} -R_n & 0 \\ 0 & S_nR_n^{-1}S_n^* - T_n \end{bmatrix} \begin{bmatrix} I_q & -R_n^{-1}S_n^* \\ 0 & I_p \end{bmatrix}. \tag{6.8}$$

Since $\rho_{d,n}(\zeta,\theta) = R_n^{-1}$ and $\rho_{g,n}(\zeta,\theta) = S_n R_n^{-1} S_n^* - T_n$, we find from (6.8) that

$$\det W_n = (-1)^q \det \rho_{d,n}^{-1}(\zeta,\theta) \det \rho_{g,n}(\zeta,\theta). \tag{6.9}$$

According to (5.3),

$$W_n = B_n^{-1}(\zeta) j B_n^{*-1}(\zeta), \qquad B_n(\zeta) = \overset{\curvearrowleft}{\prod_{k=0}^{n}} b_k(\zeta), \tag{6.10}$$

and it follows from (4.8) that $b_k(\zeta) = I + (1-\zeta)P_k/\zeta$, where P_k is expressed in terms of the corresponding Schur parameter $c_0^{(k)}$ by (2.8) and (2.9), and satisfies (2.10). Using the fact that

$$b^{-1}(\zeta) = j b_k^*(1/\hat{\zeta}) j = I - (1-\zeta)P_k,$$

we obtain

$$b_k^{-1}(\zeta) j b_k^{*-1}(\zeta) = (I - (1 - |\zeta|^2)P_k)j.$$

Hence it follows from (6.10) and the properties of P_k, $k = 0, 1, 2, \dots, n$, that

$$\begin{aligned} \det W_n &= (\det j) \prod_{k=0}^{n} \det(I - (1-|\zeta|^2)P_k) \\ &= (\det j) \prod_{k=0}^{n} |\zeta|^{2p} = (-1)^q |\zeta|^{2p(n+1)}. \end{aligned}$$

Hence, by taking account of (6.9), we obtain

$$\begin{aligned} \det \rho_{d,n}(\zeta,\theta) &= (-1)^q (\det W_n)^{-1} \det \rho_{g,n}(\zeta,\theta) \\ &= \frac{1}{|\zeta|^{2p(n+1)}} \det \rho_{g,n}(\zeta,\theta) = \det r_{g,n}(\zeta,\theta), \end{aligned}$$

as was to be proved.

COROLLARY. *We have the equations*

$$\det \rho_{d,\infty}(\zeta,\theta) = \det r_{g,\infty}(\zeta,\theta) = \det \rho_{d,\infty}(\bar{\zeta},\theta). \tag{6.11}$$

It follows from Orlov's fundamental theorem [3] that, in particular, $\operatorname{rank} \rho_{d,\infty}(\zeta,\theta)$, and therefore also $\operatorname{rank} \rho_{d,\infty}(\zeta,\hat{\theta})$, is independent of the choice of the point $\zeta \in D$. Therefore $\operatorname{rank} \rho_{d,\infty}(\zeta,\theta)$ and $\operatorname{rank} r_{g,\infty}(\zeta,\theta)$ are constants for $\zeta \in D$, and in computing them the point $\zeta \in D$ can be chosen arbitrarily. Consequently there is the possibility of classifying "infinite" Schur problems, and at the same time the functions $\theta(\zeta) \in S_{p,q}$, by means of the values of the ranks of the limit radii $\rho_{d,\infty}(\zeta,\theta)$ and $r_{g,\infty}(\zeta,\theta)$. The rest of the paper is devoted to investigating this classification.

In this connection, we note that the following proposition follows from (6.11).

THEOREM 6.3. *If one of the limit radii $\rho_{d,\infty}(\zeta,\theta)$ or $r_{g,\infty}(\zeta,\theta)$ has full rank, then the other has the same property.*

2. We consider the case $p = q$. Be repeating the discussion given in [4], we can obtain the following two propositions which play a fundamental role in what follows.

THEOREM 6.4. *Let there be given a product*

$$B_n^{-1}(\zeta) = \prod_{k=0}^{n} b_k^{-1}(\zeta) = \prod_{k=0}^{n} (I - (1-\zeta)P_k),$$

of finitely many binomial j-expansive factors of full rank, and let

$$W_k = B_k^{-1}(\zeta_0) j B_k^{*-1}(\zeta_0) = \begin{pmatrix} -R_k & S_k^* \\ S_k & -T_k \end{pmatrix}, \quad k = 0, 1, \dots, n.$$

We define the successive matrices r_0, u_0, v_0; τ_1, r_1, u_1, v_1; $\tau_2, r_2, u_2, v_2; \dots$; τ_n, r_n, u_n, v_n by means of the following recurrent process:

$$P_0 = \begin{bmatrix} \xi_0^* \\ \eta_0^* \end{bmatrix} [\xi_0, \eta_0] j; \qquad r_0 = I + (1 - |\zeta_0|^2)\xi_0\xi_0^*,$$

$$\xi_0 = (\xi_0\xi_0^*)^{1/2} u_0, \qquad \eta_0 = (\eta_0\eta_0^*)^{1/2} v_0;$$

$$\tau_k = \begin{pmatrix} r_{k-1}^{-1/2} u_{k-1} r_{k-2}^{-1/2} u_{k-2} \cdots r_0^{-1/2} u_0 & 0 \\ 0 & |\zeta_0|^{-k} r_{k-1}^{1/2} v_{k-1} r_{k-2}^{1/2} v_{k-2} \cdots r_0^{1/2} v_0 \end{pmatrix} \times \begin{pmatrix} I & 0 \\ S_{k-1} & R_{k-1}^{-1} I \end{pmatrix} B_{k-1}^{-1}(\zeta_0);$$

$$\tau_k P_k \tau_k^{-1} = \begin{bmatrix} \xi_k^* \\ \eta_k^* \end{bmatrix} [\xi_k, \eta_k] j; \qquad r_k = I + (1 - |\zeta_0|^2)\xi_k\xi_k^*;$$

$$\xi_k = (\xi_k\xi_k^*) u_k, \quad \eta_k = (\eta_k\eta_k^*)^2 v_k, \qquad 1 \le k \le n,$$

where the matrices u_k, v_k, $0 \le k \le n$, are assumed to be unitary. Then the radii of the Weyl disk K_n can be found from the formulas

$$\rho_{d,n}(\zeta_0) = u_0^* r_0^{-1/2} u_1^* r_1^{-1/2} \cdots u_n^* r_n^{-1} u_n \cdots r_1^{-1/2} u_1 r_0^{-1/2} u_0; \tag{6.12}$$

$$\rho_{g,n}(\zeta_0) = |\zeta_0|^{2n+2} v_0^* r_0^{-1/2} v_1^* r_1^{-1/2} \cdots v_n^* r_n^{-1} v_n \cdots r_1^{-1/2} v_1 r_0^{-1/2} v_0; \tag{6.13}$$

and consequently

$$r_{g,n}(\zeta_0) = v_0^* r_0^{-1/2} v_1^* r_1^{-1/2} \cdots v_n^* r_n^{-1} v_n \cdots r_1^{-1/2} v_1 r_0^{-1/2} v_0. \tag{6.14}$$

The following theorem amplifies Theorem 6.4.

THEOREM 6.5. *Let $\zeta_0 \in D$. For an arbitrary set of Hermitian matrices $r_k \ge I$, $0 \le k \le n$, and unitary matrices u_k, v_k, $0 \le k \le n$, there is a Schur problem for which the right and left radii of the Weyl disk $K_n(\zeta_0)$ are determined by* (6.12) *and* (6.13).

If we use this theorem, then as in the Nevanlinna–Pick problem [4] we can prove the following proposition.

THEOREM 6.6. *Let α and β be integers with $0 \le \alpha,\ \beta \le p-1$. Then there exists $\theta(\zeta) \in S_{p,p}$ for which* $\operatorname{rank} r_{g,\infty}(\zeta) = \alpha$ *and* $\operatorname{rank} \rho_{d,\infty}(\zeta) = \beta$.

Evidently Theorems 6.3 and 6.6 describe all possible situations in the case $p = q$.

3. The following result is of interest in the study of the general case.

LEMMA 6.1. *Let $\theta(\zeta) \in S_{p,q}$, and let $\omega_1(\zeta) \in S_{p,q_1}$ and $\omega_2(\zeta) \in S_{p_1,q}$, be connected with $\theta(\zeta)$ in the following way:*

$$\omega_1(\zeta) = [\theta(\zeta), 0], \qquad \omega_2(\zeta) = \begin{bmatrix} \theta(\zeta) \\ 0 \end{bmatrix}.$$

Then, for every $n = 0, 1, 2, \ldots$, we have the equations

$$\rho_{d,n}(\zeta, \omega_1) = \begin{pmatrix} \rho_{d,n}(\zeta, \theta) & 0 \\ 0 & I_{q_1 - q} \end{pmatrix}, \tag{6.15}$$

$$r_{g,n}(\zeta, \omega_1) = r_{g,n}(\zeta, \theta), \tag{6.16}$$

$$\rho_{d,n}(\zeta, \omega_2) = \rho_{d,n}(\zeta, \theta), \tag{6.17}$$

$$r_{g,n}(\zeta, \omega_2) = \begin{pmatrix} r_{g,n}(\zeta, \theta) & 0 \\ 0 & I_{p_1 - p} \end{pmatrix}. \tag{6.18}$$

It is sufficient to establish (6.15) and (6.16), since (6.17) and (6.18) follow from these by taking conjugates of $\theta(\zeta)$. Let

$$\theta(\zeta) = c_0 + c_1\zeta + c_2\zeta^2 + \cdots, \qquad \omega_1(\zeta) = d_0 + d_1\zeta + d_2\zeta^2 + \cdots.$$

Evidently

$$d_k = [c_k, 0], \quad d_k^* d_k = \begin{pmatrix} c_k^* c_k & 0 \\ 0 & 0 \end{pmatrix}, \quad d_k d_k^* = c_k c_k^*, \qquad k = 0, 1, 2, \ldots.$$

Therefore (6.15) and (6.16) follow respectively from (6.2) and (5.16).

THEOREM 6.7. *For $\omega(\zeta) \in S_{p,r}$ to have the form $\omega(\zeta) = [\theta(\zeta), 0]$, where $\theta(\zeta) \in S_{p,q}$, it is necessary that for every $\zeta \in D$, and sufficient that for some $\zeta \in D$, the equations*

$$\rho_{d,n}(\zeta, \omega) = \begin{pmatrix} * & 0 \\ 0 & I_{r-q} \end{pmatrix}, \qquad n = 0, 1, 2, \ldots, \tag{6.19}$$

are satisfied.

The necessity follows from Lemma 6.1. Let us turn to the proof of the sufficiency. Let $\omega(\zeta) = d_0 + d_1\zeta + d_2\zeta^2 + \cdots$,

$$D_n = \begin{bmatrix} d_0 & & & \\ d_1 & d_0 & & 0 \\ \vdots & \vdots & \ddots & \\ d_n & d_{n-1} & \ldots & d_0 \end{bmatrix}, \qquad n = 0, 1, 2, \ldots. \tag{6.20}$$

Then, according to (6.1),

$$\rho_{d,n}(\zeta, \omega) = I_r + (1 - |\zeta|^2)\Lambda_{r,n}(\zeta) D_n^* (I - D_n D_n^*)^{-1} D_n \Lambda_{r,n}^*(\zeta). \tag{6.21}$$

Let the equations (6.19) be satisfied at the point $\zeta_0 \in D$. Then it follows from (6.21) that

$$\Lambda_{r,n}(\zeta_0)D_n^*(I - D_nD_n^*)^{-1}D_n\Lambda_{r,n}^*(\zeta_0) = \begin{pmatrix} * & 0 \\ 0 & 0_{r-q} \end{pmatrix}, \qquad n = 0, 1, 2, \ldots.$$

Hence, recalling that $(I - D_nD_n^*)^{-1} \geq I$, we obtain

$$\Lambda_{r,n}(\zeta_0)D_n^*D_n\Lambda_{r,n}^*(\zeta_0) = \begin{pmatrix} * & 0 \\ 0 & 0_{r-q} \end{pmatrix}, \qquad n = 0, 1, 2, \ldots.$$

From the form (6.20) of D_n we obtain

$$D_n^*D_n = \begin{pmatrix} \tilde{d}_n[d_1^*, \ldots, d_n^*] & D_{n-1} \\ * & D_{n-1}^*D_{n-1} \end{pmatrix} \tag{6.22}$$

where $\tilde{d}_n = d_0^*d_0 + d_1^*d_1 + \cdots + d_n^*d_n$. We take account of the fact that (6.22) easily implies

$$\tilde{d}_n = \begin{pmatrix} * & 0 \\ 0 & 0_{r-q} \end{pmatrix}, \qquad n = 0, 1, \ldots.$$

Therefore

$$d_n^*d_n = \begin{pmatrix} * & 0 \\ 0 & 0_{r-q} \end{pmatrix}, \qquad n = 0, 1, 2, \ldots.$$

Consequently $d_n = [c_n, 0]$, $c_n \in L_{p,q}$, $n = 0, 1, 2, \ldots$. Consequently $\omega(\zeta) = [\theta(\zeta), 0]$, where $\theta(\zeta) \in S_{p,q}$ and $\theta(\zeta) = c_0 + c_1\zeta + c_2\zeta^2 + \cdots$, as was to be proved.

Similarly we can prove the next theorem.

THEOREM 6.8. *For $\omega(\zeta) \in S_{r,q}$ to have the form*

$$\omega(\zeta) = \begin{bmatrix} \theta(\zeta) \\ 0 \end{bmatrix}, \qquad \text{where } \theta(\zeta) \in S_{p,q},$$

it is necessary that for every $\zeta \in D$, and sufficient that for some $\zeta \in D$,

$$r_{g,n}(\zeta, \omega) = \begin{pmatrix} * & 0 \\ 0 & I_{r-p} \end{pmatrix}, \qquad n = 0, 1, 2, \ldots,$$

are satisfied.

The following theorem is one of the fundamental results of this subsection.

THEOREM 6.9. *Let p and q be positive integers. Then, for all integers α and β that satisfy the conditions $0 \leq \alpha \leq p - 1$, $0 \leq \beta \leq q - 1$, there exists $\theta(\zeta) \in S_{p,q}$ for which*

$$\operatorname{rank} r_{g,\infty}(\zeta) = \alpha, \qquad \operatorname{rank} \rho_{d,\infty}(\zeta) = \beta.$$

PROOF. If $p = q$ this proposition follows from Theorem 6.6.

Now let $p > q$, and let α and β be integers satisfying the conditions $0 \leq \alpha \leq p - 1$, $0 \leq \beta \leq q - 1$. We consider the class $S_{p,p}$. According to Theorem 6.4 the radii of the Weyl disks $K_n(\zeta_0)$ are calculated for functions of this class from formulas (6.12) and (6.13); moreover, according to Theorem 6.5

the parameters r_k, u_k, v_k, $k = 0, 1, 2, \ldots$, that appear in these formulas can be chosen arbitrarily. According to Theorem 6.6, we may choose for this parameter a function $\omega(\zeta) \in S_{p,p}$ for which

$$\operatorname{rank} r_{g,\infty}(\zeta, \omega) = \alpha, \qquad \operatorname{rank} \rho_{d,\infty}(\zeta) = \beta + p - q. \tag{6.23}$$

Notice that in the choice of $\omega(\zeta)$ the parameters r_k, u_k, $k = 0, 1, 2, \ldots$, can be chosen in the following way:

$$r_k = \begin{pmatrix} \hat{r}_k & 0 \\ 0 & I_{p-q} \end{pmatrix}, \qquad u_k = \begin{pmatrix} \hat{u}_k & 0 \\ 0 & I_{p-q} \end{pmatrix}.$$

Hence, according to (6.12), we have

$$\rho_{d,n}(\zeta_0, \omega) = \begin{pmatrix} * & 0 \\ 0 & I_{p-q} \end{pmatrix}, \qquad n = 0, 1, 2, \ldots.$$

But then it follows from Theorem 6.7 that $\omega(\zeta) = [\theta(\zeta), 0]$, $\theta(\zeta) \in S_{p,q}$. In the present case, (6.15) and (6.16) yield

$$\rho_{d,n}(\zeta, \omega) = \begin{pmatrix} \rho_{d,n}(\zeta, \theta) & 0 \\ 0 & I_{p-q} \end{pmatrix}, \qquad r_{g,n}(\zeta, \omega) = r_{g,n}(\zeta, \theta).$$

Taking limits in these equations as $n \to \infty$, and using (6.3), we obtain

$$\operatorname{rank} r_{g,\infty}(\zeta, \theta) = \alpha, \qquad \operatorname{rank} \rho_{d,\infty}(\zeta, \theta) = \beta.$$

Therefore the theorem is established in the case $p > q$. We notice that the case $p < q$ reduces to the case $p > q$ by applying the conjugation operator to the function $\theta(\zeta)$.

Clearly Theorems 6.3 and 6.9 cover all possible cases.

REMARK. It follows from our discussion that the Weyl disk $K_n(\zeta)$ is conveniently written in the form

$$C_n(\zeta) + |\zeta|^{n+1} r_{g,n}^{1/2}(\zeta) u \rho_{d,n}^{1/2}(\zeta), \qquad uu^* \leq I, \ u \in L_{p,q}.$$

Here we introduced the factor $|\zeta|^{n+1}$, on account of which the disk $K_n(\zeta)$ shrinks to a point. By (6.4), this notation is invariant under taking conjugates.

DEFINITION. We call the numbers

$$\delta_\theta = \operatorname{rank} \rho_{d,\infty}(\zeta, \theta), \tag{6.24}$$

$$\delta_{\theta^*} = \operatorname{rank} \rho_{d,\infty}(\zeta, \hat{\theta}) \tag{6.25}$$

the *deficiencies* of the contractive matrix-valued functions $\theta(\zeta)$.

In later papers we will clarify the role played by the deficiencies in the study of contractive matrix-valued functions and the corresponding contraction operators in Hilbert space.

REFERENCES

1. V. K. Dubovoĭ, *Indefinite metric in Schur's interpolation problem for analytic functions*. I, Teor. Funktsiĭ, Funktsional. Anal. i Prilozhen. Vyp. 37 (1982), 14–26; English transl. in this volume.

2. I. V. Kovalishina, *Analytic theory of a class of interpolation problems*, Izv. Akad. Nauk SSSR Ser. Mat. **47** (1983), 455–497; English transl. in Math. USSR Izv. **22** (1984).

3. S. A. Orlov, *Nested matrix disks analytically depending on a parameter, and theorems on the invariance of the ranks of radii of limit matrix disks*, Izv. Akad. Nauk SSSR Ser. Mat. **40** (1976), 593–644; English transl. in Math. USSR Izv. **10** (1976).

4. I. V. Kovalishina and V. P. Potapov, *Radii of the Weyl disks in the matrix Nevanlinna-Pick problem*, Theory of Operators in Function Spaces and Its Applications (V. A. Marchenko, editor), "Naukova Dumka", Kiev, 1981, pp. 25–49; English transl. in Amer. Math. Soc. Transl. (2) **138** (1988).

Translated by RALPH P. BOAS

Amer. Math. Soc. Transl.
(2) Vol. **144**, 1989

On Uniform and Tangential Approximation by Meromorphic Functions

A. A. NERSESYAN

A closed set E is called a *set of uniform approximation by meromorphic functions* if for any function $f(z)$ continuous on E and holomorphic in its interior E^0 and for each $\varepsilon > 0$, there exists a meromorphic function $\mu(z)$ such that $|f(z) - \mu(z)| < \varepsilon$ for $z \in E$. If, instead, for each *function* $\varepsilon(x) > 0$, continuous on $[0, \infty)$ and tending to zero arbitrarily fast as $x \to \infty$, there exists a meromorphic function $\mu(z)$ such that $|f(z) - \mu(z)| < \varepsilon(|z|)$ for $z \in E$, then the set E is called a *set of tangential approximation by meromorphic functions.*

In this paper we consider the question of finding necessary and sufficient conditions on a set in order that it be a set of uniform approximation by meromorphic functions. Necessary and sufficient conditions for tangential approximation are also found for sets having no interior.

We have the following theorems.

THEOREM 1. *A necessary and sufficient condition in order for a closed set E to be a set of uniform approximation by meromorphic functions is that for each closed disc $\overline{D}$, the set $E \cap \overline{D}$ be a set of uniform approximation by rational functions.*

THEOREM 2. *A necessary and sufficient condition in order for a closed set E, with $E^0 = \varnothing$, to be a set of tangential approximation by meromorphic functions is that for each closed disc $\overline{D}$, the set $E \cap \overline{D}$ be a set of rational approximation.*

The necessity in Theorem 1 is proved in terms of continuous analytic capacity, using the celebrated theorem of A. G. Vitushkin on rational approximation [2], [3]. The sufficiency is proved with the help of the method

1980 *Mathematics Subject Classification* (1985 *Revision*). Primary 30E10.
Translation of Izv. Akad. Nauk Armyan. SSR Ser. Mat. 7 (1972), 405–412; MR **51** #3460.

developed by M. V. Keldysh (see [1]). Theorem 2 is obtained as a corollary of Theorem 1.

We state without proof a theorem of Vitushkin [2], [3], which we shall use in the sequel.

THEOREM 3. *Let U be a bounded open set. Suppose a function $f(z)$, continuous on the Riemann sphere and holomorphic in the complement of U, can be uniformly approximated on some compact set F by rational functions whose poles lie outside of F. Then, for each $\varepsilon > 0$, there exists a rational function $r(z)$ such that $|f(z) - r(z)| < \varepsilon$ for $z \in F \cup CU$.*

According to another theorem of Vitushkin, on uniform approximation by rational functions, in order to prove the necessity in Theorem 1, it is sufficient to show that $\alpha(K \setminus E) = \alpha(K \setminus E^0)$ for any open disc K, where $\alpha(A)$ denotes the continuous analytic capacity of a set A.

We denote by $C(K, m)$ the class of functions continuous on the Riemann sphere, holomorphic outside of K, vanishing at infinity, and bounded in modulus by m.

Before proceeding to the proof of the theorem, we remark that the theorem is true in a more general form. Given a plane domain D ($\partial D \neq \varnothing$), we seek necessary and sufficient conditions on a set $E \subset D$, relatively closed in D, for E to be a set of uniform (respectively tangential) approximation by functions meromorphic in D. In this situation, Theorems 1 and 2 can be formulated in the following manner:

THEOREM 1* (2*). *A relatively closed subset $E \subset D$ is a set of uniform (respectively, tangential in case $E^0 = \varnothing$) approximation by functions meromorphic in D if and only if $E \cap K$ is a set of rational approximation for each closed disc $K \subset D$.*

We shall prove the theorem for $D = C$, restricting ourselves to remarks concerning the general case.

Let us prove the necessity of the condition in Theorem 1. Let E be a set of uniform approximation by meromorphic functions, K an open disc, and $\varepsilon > 0$ arbitrary but fixed. From the definition of continuous analytic capacity, there exist a closed set $F \subset K \setminus E^0$ and a function $f \in C(F, 1)$ such that $f'(\infty) > \alpha(K \setminus E^0) - \varepsilon$. Clearly, f is holomorphic in some neighborhood of CK, and since f is holomorphic on E^0 and continuous on E, f may be uniformly approximated on E by meromorphic functions. This means that f can be uniformly approximated on $E \cap \overline{K}$ by rational functions. From this we obtain, in view of Theorem 3, the existence of a sequence $\{f_n\}$ of rational functions for which $f_n \to f$ uniformly on $E \cup CK$. After appropriate modifications of these functions on $K \setminus E$, we may suppose, without loss of generality, that $f_n \in C(K \setminus E, 1)$. Then, $f_n'(\infty) \to f'(\infty)$ and $|f_n'(\infty)| \leq \alpha(K \setminus E)$. Thus, $\alpha(K \setminus E^0) - \varepsilon < f'(\infty) < \alpha(K \setminus E)$. Since ε was arbitrary, we have $\alpha(K \setminus E^0) \leq \alpha(K \setminus E)$. The opposite inequality is obvious.

All of the calculations in the following proof of sufficiency are valid also in the general case, if instead of discs (centered at the origin) we take the sets $D \setminus V_\sigma(\partial D)$ for appropriate $\sigma > 0$ ($V_\sigma(\partial D)$ denotes an open spherical σ-neighborhood of ∂D) and remark that from Vitushkin's theorem easily follows the equivalence of the condition in Theorem 1 to the requirement that $E \setminus V_\sigma$ be a set of rational approximation.

Let f be a function continuous on E and holomorphic on E^0, and let ε be a positive constant. Let α be a positive, continuous, monotone function on $[0, \infty)$ such that

$$0 < \alpha(x) < \varepsilon \quad \text{and} \quad \iint_{|\zeta|<\infty} \frac{\alpha(|\zeta| - \frac{1}{2})}{|\zeta - z|}\, d\xi\, d\eta < \varepsilon \tag{1}$$

for all z, where $\zeta = \xi + i\eta$.

Under our assumptions on E, there exists a sequence $\{\varphi_n\}$ of rational functions, with poles outside of E, satisfying the following inequalities:

$$\begin{aligned} |\varphi_1(z) - f(z)| &< \alpha(2 + \tfrac{1}{2}), \qquad z \in E \cap \{|\zeta| < 2 + \tfrac{1}{2}\}, \\ |\varphi_n(z) - f(z)| &< \alpha(2n + \tfrac{1}{2}), \qquad z \in E \cap \{2(n-1) - \tfrac{1}{2} \le |\zeta| \le 2n + \tfrac{1}{2}\}, \quad (*) \\ & \qquad n = 2, 3, \ldots. \end{aligned}$$

With the help of this sequence, we define a function $\varphi(z, r)$ in the following way: for fixed z, $\varphi(z, r) = \varphi_n(z)$ if $2(n-1) \le r < 2n$, $n = 1, 2, \ldots$.

We have $|\varphi(z, r) - f(z)| < \alpha(r)$ for $r - \frac{1}{2} \le |z| \le r + \frac{1}{2}$.

By definition, the function $\varphi(z, r)$ is piecewise constant in r for fixed z, and an arbitrary point z falls in at most two distinct annuli, on whose intersection with E, $(*)$ is valid.

For each point $z \in E$, there exists an open disc with center z and radius $\delta(z) > 0$ such that the variation of the function $\varphi(z, r)$, approximating $f(z)$ at this point, does not exceed $\alpha(r' - 1/2)$, where r' is the largest radius of the annulus corresponding to this function. The collection of these discs is an open covering of the set E. From this covering we may extract a locally finite subcover. The union E^* of the discs of this subcover contains the closed set E. The boundary ∂E^* of E^* consists of countably many Jordan curves or points made up of a locally finite family of circular arcs and points. We may (and shall) choose our subcover so that no boundary component of E^* is a point. Let $\{\gamma_k\}$, $k = 1, 2, \ldots$, be the sequence of components of ∂E^*.

On E we define the function

$$\varphi(z) = \int_{|z|-1/2}^{|z|+1/2} \varphi(z, t)\, dt.$$

For $z \in E$, we have $|f(z) - \varphi(z)| < \varepsilon$.

We define a sequence of functions $\{Q_n\}$ in the following way:

$$Q_n(z) = \frac{1}{2\pi i} \sum_{l=1}^{n} \left(\sum_{k=1}^{\infty} \int_{\gamma_k \cap (C_l \setminus C_{l-1})} \frac{\varphi(\zeta)}{\zeta - z}\, d\zeta \right) + \frac{1}{2\pi i} \int_{L_n} \frac{\varphi(\zeta)}{\zeta - z}\, d\zeta,$$

where $C_l = (|z| \leq l)$, $C_0 = \varnothing$, and $L_n = E^* \cap (|z| = n)$.

Corresponding to this sequence, we define functions

$$R_n(z) = \sum_{l=1}^{n} r_l(z) + \frac{1}{2\pi i} \int_{L_n} \frac{\varphi(\zeta)}{\zeta - z} \, d\zeta, \qquad n = 1, 2, \dots,$$

where r_l is a rational function with poles outside of $E \cup C_{l-1}$, such that

$$\text{(2)} \qquad \left| r_l(z) - \frac{1}{2\pi i} \sum_{k=1}^{\infty} \int_{\gamma_k \cap (C_l \setminus C_{l-1})} \frac{\varphi(\zeta)}{\zeta - z} \, d\zeta \right| < \frac{\varepsilon}{2^l},$$
$$z \in E \cup C_{l-2}, \quad C_0 = C_{-1} = \varnothing, \quad l = 1, 2, \dots.$$

We find the functions r_l by replacing the integrals over a finite number of arcs by suitable integral sums.

We show that the sequence of functions $\{R_n\}$ converges uniformly on each disc S. To this end, consider the difference

$$\frac{1}{2\pi i} \int_{L_{n+m}} \frac{\varphi(\zeta)}{\zeta - z} \, d\zeta - \frac{1}{2\pi i} \int_{L_n} \frac{\varphi(\zeta)}{\zeta - z} \, d\zeta,$$

where n is such that $\overline{S} \subset (|z| < n - 2)$, and $m > 0$ is an integer.

Notice that if $L_r = E^* \cap (|\zeta| = r)$ and ψ is holomorphic on $\overline{E}^*$, then by Cauchy's theorem,

$$d_r \int_{L_r} \psi(\zeta) \, d\zeta = \sum_{(r)} \psi(\zeta) \, d\zeta.$$

The sum on the right side is over the points of intersection of $\{|\zeta| = r\}$ with ∂E^*. By $d\zeta$ we denote the differential of the arc along ∂E^* corresponding to a negative circuit of E^*.

Because of this, for $z \in S$ and $r > n$ we have

$$\begin{aligned}
d_r \int_{L_r} \frac{\varphi(\zeta)}{\zeta - z} \, d\zeta &= d_r \int_{r-1/2}^{r+1/2} dt \int_{L_r} \frac{\varphi(\zeta, t)}{\zeta - z} \, d\zeta \\
&= \left\{ \int_{L_r} \frac{\varphi(\zeta, t)}{\zeta - z} \, d\zeta \right\}_{t=r-1/2}^{t=r+1/2} d_r + \int_{r-1/2}^{r+1/2} d_r \left\{ \int_{L_r} \frac{\varphi(\zeta, t)}{\zeta - z} \, d\zeta \right\} dt \\
&= \left\{ \int_{L_r} \frac{\varphi(\zeta, r + \frac{1}{2}) - \varphi(\zeta, r - \frac{1}{2})}{\zeta - z} \, d\zeta \right\} d_r \\
&\qquad + \int_{r-1/2}^{r+1/2} \left\{ \sum_{(r)} \frac{\varphi(\zeta, t)}{\zeta - z} \, d\zeta \right\} dt \\
&= \left\{ \int_{L_r} \frac{\varphi(\zeta, r + \frac{1}{2}) - \varphi(\zeta, r - \frac{1}{2})}{\zeta - z} \, d\zeta \right\} d_r + \sum_{(r)} \frac{\varphi(\zeta)}{\zeta - z} \, d\zeta.
\end{aligned}$$

Integrating this equation with limits $n \le t \le n+m$, we obtain

$$\begin{aligned}
&\int_{L_{n+m}} \frac{\varphi(\zeta)}{\zeta - z}\, d\zeta - \int_{L_n} \frac{\varphi(\zeta)}{\zeta - z}\, d\zeta \\
&\quad = \int_n^{n+m} dr \int_{L_r} \frac{\varphi(\zeta, r+\frac{1}{2}) - \varphi(\zeta, r-\frac{1}{2})}{\zeta - z}\, d\zeta \\
&\qquad - \sum_{l=n+1}^{n+m} \left(\sum_{k=1}^{\infty} \int_{\gamma_k \cap (C_l \setminus C_{l-1})} \frac{\varphi(\zeta)}{\zeta - z}\, d\zeta \right) \\
&\quad = \iint_{E^* \cap (C_{n+m} \setminus C_n)} \frac{\varphi(\zeta, r+\frac{1}{2}) - \varphi(\zeta, r-\frac{1}{2})}{\zeta - z} \frac{d\zeta}{|d\zeta|}\, d\xi\, d\eta \\
&\qquad - \sum_{l=n+1}^{n+m} \left(\sum_{k=1}^{\infty} \int_{\gamma_k \cap (C_l \setminus C_{l-1})} \frac{\varphi(\zeta)}{\zeta - z}\, d\zeta \right).
\end{aligned} \tag{3}$$

In equation (3), the integrals under the sums are taken in the positive direction around E^*.

According to (2), (3), and the choice of the number n for S, we have

$$\begin{aligned}
|R_{n+m}(z) - R_n(z)| &= \left| \sum_{l=n+1}^{n+m} r_l(z) + \frac{1}{2\pi i} \int_{L_{n+m}} \frac{\varphi(\zeta)}{\zeta - z}\, d\zeta - \frac{1}{2\pi i} \int_{L_n} \frac{\varphi(\zeta)}{\zeta - z}\, d\zeta \right| \\
&= \left| \sum_{l=n+1}^{n+m} r_l(z) - \sum_{l=n+1}^{n+m} \left(\sum_{k=1}^{\infty} \frac{1}{2\pi i} \int_{\gamma_k \cap (C_l \setminus C_{l-1})} \frac{\varphi(\zeta)}{\zeta - z}\, d\zeta \right) \right. \\
&\quad + \sum_{l=n+1}^{n+m} \left(\sum_{k=1}^{\infty} \frac{1}{2\pi i} \int_{\gamma_k \cap (C_l \setminus C_{l-1})} \frac{\varphi(\zeta)}{\zeta - z}\, d\zeta \right) \\
&\quad \left. + \frac{1}{2\pi i} \int_{L_{n+m}} \frac{\varphi(\zeta)}{\zeta - z}\, d\zeta - \frac{1}{2\pi i} \int_{L_n} \frac{\varphi(\zeta)}{\zeta - z}\, d\zeta \right| \\
&\le \varepsilon \sum_{l=n+1}^{n+m} \frac{1}{2^l} \\
&\quad + \left| \frac{1}{2\pi} \iint_{E^* \cap (C_{n+m} \setminus C_n)} \frac{\varphi(\zeta, r+\frac{1}{2}) - \varphi(\zeta, r-\frac{1}{2})}{\zeta - z} \frac{d\zeta}{|d\zeta|}\, d\xi\, d\eta \right| \\
&\le \varepsilon \sum_{l=n+1}^{n+m} \frac{1}{2^l} + \frac{2}{\pi} \iint_{C_{n+m} \setminus C_n} \frac{\alpha(|\zeta| - \frac{1}{2})}{|\zeta - z|}\, d\xi\, d\eta.
\end{aligned}$$

By (1), the right side of this inequality can be made as small as we wish by choosing n sufficiently large.

Thus, we have shown that the series $R_1(z) + \sum_1^\infty (R_{n+1}(z) - R_n(z))$ converges uniformly on any finite disc, if we discard a finite number of initial terms. Since the singularities of these discarded terms can only be a finite number of poles, the function

$$\mu(z) = R_1(z) + \sum_{n=1}^{\infty} (R_{n+1}(z) - R_n(z))$$

is meromorphic.

Fix $z \in E$ and let $p > 0$ be such that $(|\zeta - z| \le p) \subset E^0$.

In a manner similar to the proof of the preceding equations, and applying the corresponding calculations to the set $E^* \setminus (|\zeta - z| < p)$, we obtain

$$
(4)\qquad Q_n(z) - \frac{1}{2\pi i}\int_{|\zeta - z| = p} \frac{\varphi(\zeta)}{\zeta - z}\, d\zeta
= \frac{1}{2\pi i}\iint_{(E^*\cap C_n)\setminus(|\zeta - z|<p)} \frac{\varphi(\zeta, |\zeta| + \frac{1}{2}) - \varphi(\zeta, |\zeta| - \frac{1}{2})}{\zeta - z}\,\frac{d\zeta}{|d\zeta|}\, d\xi\, d\eta.
$$

Choose $N > 0$ such that $|\sum_N^\infty (R_{n+1}(z) - R_n(z))| < \varepsilon$ for $z \in E$.

According to (1), (2), (4), and the construction of the set E^*, for sufficiently small p we have

$$
\begin{aligned}
|\mu(z) - \varphi(z)| &\le \left|\sum_{n=N}^{\infty} (R_{n+1}(z) - R_n(z))\right| + |R_N(z) - \varphi(z)| \\
&< \varepsilon + |R_N(z) - \varphi(z)| \\
&< \varepsilon + (1 - 2^{-N})\varepsilon + |Q_N(z) - \varphi(z)| \\
&< 2\varepsilon + \left| Q_N(z) - \frac{1}{2\pi i}\int_{|\zeta - z| = p} \frac{\varphi(\zeta)}{\zeta - z}\, d\zeta + \frac{1}{2\pi i}\int_{|\zeta - z| = p} \frac{\varphi(\zeta) - \varphi(z)}{\zeta - z}\, d\zeta \right| \\
&< 2\varepsilon + \left| \frac{1}{2\pi}\iint_{(E^*\cap C_N)\setminus(|\zeta - z|<p)} \frac{\varphi(\zeta, |\zeta| + \frac{1}{2}) - \varphi(\zeta, |\zeta| - \frac{1}{2})}{\zeta - z}\,\frac{d\zeta}{|d\zeta|}\, d\xi\, d\eta \right| \\
&\quad + \frac{1}{2\pi}\left| \int_{|\zeta - z| = p} \frac{\varphi(\zeta) - \varphi(z)}{\zeta - z}\, d\zeta \right| \\
&\le 2\varepsilon + \frac{2}{\pi}\iint_{|\zeta| < \infty} \frac{\alpha(|\zeta| - \frac{1}{2})}{|\zeta - z|}\, d\xi\, d\eta + \max_{\zeta, |\zeta - z| = \rho} |\varphi(\zeta) - \varphi(z)| \\
&< 4\varepsilon.
\end{aligned}
$$

From this, bearing in mind that $\varphi(z)$ approximates $f(z)$ on E within ε, we obtain $|\mu(z) - f(z)| < 5\varepsilon$. This completes the proof.

The following generalization of Runge's theorem can be proved in a manner analogous to the proof of Theorem 1: *If a function f is holomorphic on a closed set E, then for each $\varepsilon > 0$ we can find a function μ, meromorphic on the finite plane, such that $|f(z) - \mu(z)| < \varepsilon$ for $z \in E$.*

Suppose, now, that $E^0 = \varnothing$, f is continuous on E, and $\varepsilon(x) > 0$ is an arbitrary positive continuous function on $[0, \infty)$ which tends to zero as $x \to \infty$. Without loss of generality, we may assume that $\varepsilon(x) \le 1$. By Theorem 1, there exists a meromorphic function $\mu(z)$ such that

$$
\left| \frac{2}{\varepsilon(|z|)} - \mu(z) \right| < 1 \qquad \text{for } z \in E.
$$

Then

$$
|\mu(z)| > \frac{2}{\varepsilon(|z|)} - 1 = \frac{2 - \varepsilon(|z|)}{\varepsilon(|z|)} \ge \frac{1}{\varepsilon(|z|)},
$$

and consequently, for $z \in E$,

$$\frac{1}{|\mu(z)|} < \varepsilon(|z|).$$

Applying Theorem 1 to the function $f(z)\mu(z)$, we obtain the existence of a function ν such that $|f(z)\mu(z) - \nu(z)| < 1$ for $z \in E$. Thus,

$$\left| f(z) - \frac{\nu(z)}{\mu(z)} \right| < \frac{1}{|\mu(z)|} < \varepsilon(|z|).$$

This proves Theorem 2.

In conclusion, it is a pleasure for the author to express his deepest gratitude to Dr. N. U. Arakelyan for his assistance during the course of this work.

REFERENCES

1. S. N. Mergelyan, *Uniform approximation to functions of a complex variable*, Uspekhi Mat. Nauk **7** (1952), no. 2(48), 31–122; English transl. in Amer. Math. Soc. Transl. (1) **3** (1962).

2. A. G. Vitushkin, *Analytic capacity of sets in problems of approximation theory*, Uspekhi Mat. Nauk **22** (1967), no. 6 (138), 141–199; English transl. in Russian Math. Surveys **22** (1967).

3. Lawrence Zalcman, *Analytic capacity and rational approximation*, Lecture Notes in Math., vol. 50, Springer-Verlag, 1968.

Translated by P. M. GAUTHIER

Amer. Math. Soc. Transl.
(2) Vol. **144**, 1989

On Entire Functions of Completely Regular Growth in Several Variables

UDC 517.55

S. YU. FAVOROV

According to the definition of Levin and Pfluger [1], an entire function $f(w)$ for $w \in \mathbf{C}$ of normal type relative to a proximate order $\rho(t)$ is called a function of *completely regular growth* if for all $\varphi \in [0, 2\pi)$ we have

$$\lim_{t\to\infty}{}^{*} \frac{\ln|f(te^{i\varphi})|}{t^{\rho(t)}} = h_f(\varphi),$$

where $h_f(\varphi)$ is the indicator for the function $f(w)$, and * means that t does not pass through a certain set E_φ such that the Lebesgue measure of $E_\varphi \cap [0, R]$ is $o(R)$ as $R \to \infty$.

As has been noted, regular growth for an entire function is equivalent to the regular distribution of its set of zeros.

For subharmonic functions in $\mathbf{R}^m$ a definition of completely regular growth was given by Azarin [2]. He introduced the concept of regular mass distribution, and proved that a subharmonic function in $\mathbf{R}^m$ has completely regular growth if and only if its approximating mass is regularly distributed.

In [3], where holomorphic functions of several variables are considered, it was proved in particular that for an entire function of completely regular growth $f(w)$, $w \in \mathbf{C}$, the limit of $t^{-\rho(t)} \ln|f(tw)|$ exists as $t \to \infty$ in the space $D'(\mathbf{C})$ and is equal almost everywhere to $|w|^\rho h_f(\arg w)$. Azarin [4] and Agranovich [5] have shown that the definition given above for subharmonic functions of completely regular growth is equivalent to the following.

A subharmonic function $v(x)$ for $x \in \mathbf{R}^m$ is called a *function of completely regular growth* if $\lim_{t\to\infty} t^{-\rho(t)} v(tx)$ exists in $D'(\mathbf{R}^m)$. This limit, as was noted in [4]–[6], coincides almost everywhere with the indicator $h(x) = \overline{\lim}_{t\to\infty} t^{-\rho(t)} v(tx)$.

1980 *Mathematics Subject Classification* (1985 *Revision*). Primary 32A15.
Translation of Teor. Funktsiĭ, Funktsional. Anal. i Prilozhen. Vyp. 38 (1982), 103–111; MR **84e**:32004.

Let $F(z)$ be an entire function in $\mathbf{C}^n$. We shall say that $F(z)$ has *completely regular growth in sense* (A) if the subharmonic function $\ln|F(z)|$ has completely regular growth in $\mathbf{R}^{2n}$ in the sense of the definition given above. Certain properties of functions in this class were considered in [7]. We recall that the indicator of the subharmonic function $\ln|F(z)|$ is called the *radial indicator* of the entire function $F(z)$ and is denoted by $L_r(z;F)$.

Following Gruman [8], and also [9] and [10], we shall say that an entire function $F(z)$ has *completely regular growth in sense* (B) if for almost all $z \in \mathbf{C}^n$ the function $F(z \cdot w)$ for $w \in \mathbf{C}$ is a function of completely regular growth in the plane of the variable w.

It was shown in [3], [5], and [6] that for an entire function $F(z)$, $z \in \mathbf{C}^n$, of completely regular growth in sense (B) the limit of $t^{-\rho(t)} \ln|F(tz)|$ as $t \to \infty$ exists in $D'(\mathbf{C}^n)$ and is equal to $L_r(z;F)$; consequently $F(z)$ is a function of completely regular growth in sense (A).

In this article we prove the following fact.

THEOREM. *For a fixed $\rho > 0$ there exists an entire function $F(z)$ in $\mathbf{C}^2$ of completely regular growth in sense* (A) *relative to the order ρ with the property that for almost every $z \in \mathbf{C}^2$ the function $F(z \cdot w)$, $w \in \mathbf{C}$, is not a function of completely regular growth in the plane of the variable w and consequently is not a function of completely regular growth in sense* (B).

We note that an example of a subharmonic function of completely regular growth $v(x)$, $x \in \mathbf{R}^m$, $m > 2$, for which $\lim^*_{t\to\infty} t^{-\rho} v(tx)$ does not exist for all $x \in \mathbf{R}^m$ was constructed in [11].

NOTATION. We denote points in the space $\mathbf{C}^2$ by $z = (z_1, z_2)$ and $\zeta = (\zeta_1, \zeta_2)$. Here $|z| = (|z_1|^2 + |z_2|^2)^{1/2}$. We set $B(\zeta, r) = \{z \in \mathbf{C}^2 : |z - \zeta| < r\}$, $S = \{z : |z| = 1\}$, and $\Gamma = \{z \in \mathbf{C}^2 \backslash \{0\} : \min\{|z_1||z|^{-1}, |z_2||z|^{-1}\} \geq \frac{1}{3}\}$. For $z' \in \mathbf{C}^2 \backslash \{0\}$ we denote by $P(z', a, b)$ the polydisc

$$\{z \in \mathbf{C}^2 : z = (z'_1, z'_2) + \lambda(z'_1, z'_2) + \mu(-z'_2, z'_1),\ \lambda, \mu \in \mathbf{C},\ |\lambda| < a,\ |\mu| < b\}.$$

We note that the linear transformation

$$(z_1, z_2) \to (\lambda, \mu): \quad \begin{aligned} \lambda &= |z'|^{-2}(z_1 \overline{z}'_1 + z_2 \overline{z}'_2) - 1, \\ \mu &= |z'|^{-2}(z_2 z'_1 - z_1 z'_2) \end{aligned} \tag{1}$$

maps the polydisc $P(z', a, b)$ into the polydisc $\{(\lambda, \mu) : |\lambda| < a,\ |\mu| < b\}$.

We say that a set $E \subset \mathbf{C}^2$ is a set of zero relative measure if $m(E \cap B(0, R)) = O(R^4)$ as $R \to \infty$, where m is Lebesgue measure in $\mathbf{C}^2$.

We recall that the Levi form of a locally summable function $\theta(z)$ in $\mathbf{C}^2$ is given by

$$\frac{\partial^2 \Theta(z)}{\partial z_1 \partial \overline{z}_1} |\zeta_1|^2 + 2\mathrm{Re} \frac{\partial^2 \Theta(z)}{\partial z_1 \partial \overline{z}_2} \zeta_1 \overline{\zeta}_2 + \frac{\partial^2 \Theta(z)}{\partial z_2 \partial \overline{z}_2} |\zeta_2|^2, \qquad \zeta \in \mathbf{C}^2.$$

It is well known that a continuous function is plurisubharmonic if and only if its Levi form is nonnegative for all $\zeta \in \mathbf{C}^2$.

The proof of the theorem is based on the following lemmas.

LEMMA 1. *For any* $\rho > 0$, $z' \in \Gamma$, $z \in P(z', \frac{1}{4}, \frac{1}{4})$, *and* $\zeta \in \mathbf{C}^2$, *the Levi form of the function* $|z_1|^\rho + |z_2|^\rho$ *is bounded from below by* $\sigma|z'|^{\rho-2}|\zeta|^2$, *where* $\sigma = \sigma(\rho)$ *is a positive constant.*

We omit the proof.

LEMMA 2. *For any* $\rho > 0$, $\delta \in (0, \frac{1}{12})$, *and* $z' \in \Gamma$, *there exist constants* $\varepsilon = \varepsilon(\rho) > 0$ *and* $\tau = \tau(\rho, \delta) > 0$ *and a function* $\varphi(z) = \varphi(z, z', \delta, \rho)$ *so that:* a) *the function* $(1-\varepsilon\varphi(z))(|z_1|^\rho+|z_2|^\rho)$ *is plurisubharmonic in* $\mathbf{C}^2$, b) $\operatorname{supp}\varphi \in P(z', \frac{1}{8}, \delta)$, c) $0 \le \varphi(z) \le 1$ *for all* z, *and* d) $\varphi(z) = 1$ *for* $z \in P(z', \frac{1}{12}, \tau)$.

PROOF. Let $\chi(w)$ be a function in $C^\infty(\mathbf{C})$ such that $\chi(w) = 1$ for $|w| \le 1$, $\chi(w) = 0$ for $|w| \ge \frac{3}{2}$, and $0 \le \chi(w) \le 1$ for all $w \in \mathbf{C}$. We set

$$\psi(z) = \psi(z; z') = \chi(12\lambda(z; z')) \cdot \chi(12\mu(z; z')),$$

where the functions $\lambda(z; z')$ and $\mu(z; z')$ are defined by (1). For $z \in P(z', \frac{1}{8}, \frac{1}{8})$ we have $\frac{1}{2}|z'| \le |z| \le 2|z'|$. Therefore for such z we have

$$\left|\frac{\partial\psi}{\partial z_i}\right| \le K_1|z'|^{-1}, \quad \left|\frac{\partial^2\psi}{\partial z_i \partial \overline{z}_j}\right| \le K_2|z'|^{-2}, \qquad i, j = 1, 2.$$

It follows from this and from Lemma 1 that for $\varepsilon \le \varepsilon_0(K_1, K_2, \sigma)$, where σ is the constant from Lemma 1, the Levi form of the function $[|z_1|^\rho + |z_2|^\rho][1 - \varepsilon\psi(z)]$ is nonnegative. Further, we set

$$\omega(t) = \begin{cases} 0, & t \ge \delta, \\ (t-\delta)^2, & \frac{\delta}{2} \le t < \delta, \\ \frac{\delta^2}{2}\ln\left(\frac{\delta}{2}\sqrt{e}t^{-1}\right), & 0 < t < \frac{\delta}{2}. \end{cases}$$

The function $\omega(t)$ is twice differentiable in $(0, \infty)$, and $\partial^2\omega(|\mu|)/\partial\mu\partial\overline{\mu} \le 1$. Therefore for $z \in P(z', \frac{1}{8}, \frac{1}{8})$ the Levi form of the function $\Theta(z) = \omega(|\mu(z; z')|)$ does not exceed $|z'|^{-2}|\zeta|^2$, and consequently for $z' \in \Gamma$ and $\varepsilon \le \sigma$, where σ is the constant from Lemma 1, the function

$$|z_1|^\rho + |z_2|^\rho - \varepsilon|z'|^\rho\omega(|\mu(z; z')|)$$

is plurisubharmonic in $P(z', \frac{1}{8}, \frac{1}{8})$. We choose $\tau = \tau(\rho, \delta)$ so that $|z'|^\rho\omega(\tau) \ge |z_1|^\rho + |z_2|^\rho$ in $P(z', \frac{1}{8}, \frac{1}{8})$. The function

$$\varphi(z) = \min\{\psi(z; z'); |z'|^\rho(|z_1|^\rho + |z_2|^\rho)^{-1}\omega(|\mu(z; z')|)\}$$

satisfies conditions b)–d). Since the maximum of two plurisubharmonic functions is again plurisubharmonic, condition a) is satisfied for $\varepsilon \le \min\{\sigma, \varepsilon_0\}$. This proves the lemma.

LEMMA 3. *Let* $\Delta(t)$ *be a positive function which decreases monotonically to zero as* $t \to \infty$. *There exists on* $\mathbf{C}^2$ *a plurisubharmonic function* $\Phi(z)$ *with the following properties*: a) $\Phi(z) \le |z_1|^\rho + |z_2|^\rho$; b) *the set* $A_1 \overset{\text{def}}{=} \{z \in \mathbf{C}^2 \colon \Phi(z) < |z_1|^\rho + |z_2|^\rho\}$ *is the union of a sequence of polydiscs* $P(z^{(n)}, \frac{1}{8}, \delta_n)$, *where* $|z^{(n+1)}| \ge 2|z^{(n)}|$ *and* $\Delta(|z^{(n)}|) \le \delta_n \le \frac{1}{12}$ *for all* n; *moreover* $\delta_n \to 0$ *as*

$n \to \infty$; c) *the set* A_1 *has zero relative measure*; *and* d) *for each* $\zeta \in S \cap \Gamma$ *there is a sequence* $t' \to \infty$ *so that for* $t \in [t' - \frac{t'}{15}, t' + \frac{t'}{15}]$ *and* $z \in B(t\zeta, 1)$ *we have* $\Phi(z) \le (1-\varepsilon)(|z_1|^\rho + |z_2|^\rho)$, *where* $\varepsilon = \varepsilon(\rho)$ *is the same as in Lemma* 2.

PROOF. We set $n_0 = 0$ and choose r_0 so that $r_0 \ge 1$ and $\Delta(r_0) \le \frac{1}{62}$. Let $\zeta^{(n)}$ for $n = n_0 + 1, \dots, n_1$ be the centers of balls of radius $\frac{1}{2}\tau(\rho, \Delta(r_{n_0}))$ forming a finite covering of the compact set $S \cap \Gamma$, where τ is the constant from Lemma 2. Let $r_{n_0+1} = \max\{2r_{n_0}; 2\tau^{-1}(\rho, \Delta(r_{n_0}))\}$, $r_{n+1} = 2r_n$, $n = n_0 + 1, \dots, n_1 - 1$; $z^{(n)} = r_n\zeta^{(n)}$, $\delta_n = \Delta(r_{n_0})$, and $\tau_n = \tau(\rho, \Delta(r_{n_0}))$, $n = n_0 + 1, \dots, n_1$. Further let $\zeta^{(n)}$ for $n = n_1 + 1, \dots, n_2$ be the centers of balls of radius $\frac{1}{2}\tau(\rho, \Delta(r_{n_1}))$ forming an open covering of the compact set $S \cap \Gamma$. We now set $r_{n_1+1} = \max\{2r_{n_1}; 2\tau^{-1}(\rho; \Delta(r_{n_1}))\}$, $r_{n+1} = 2r_n$, $n = n_1 + 1, \dots, n_2 - 1$; $z^{(n)} = r_n\zeta^{(n)}$, $\delta_n = \Delta(r_{n_1})$, and $\tau_n = \tau(\rho; \Delta(r_{n_1}))$, $n = n_1 + 1, \dots, n_2$. Further, let $\zeta^{(n)}$ for $n = n_2 + 1, \dots, n_3$ be the centers for balls of radius $\frac{1}{2}\tau(\rho, \Delta(r_{n_2}))$ forming a finite covering of the compact set $\Gamma \cap \rho$, etc. Thus we construct a sequence of points $\zeta^{(n)}$, $z^{(n)}$, r_n, τ_n, δ_n with the properties indicated in b) of the lemma. Here we may assume that $\zeta^{(n)} \in \Gamma \cap S$ for all n. We set

$$\Phi(z) = (|z_1|^\rho + |z_2|^\rho)[1 - \varepsilon \sum_{n=1}^{\infty} \varphi(z; z^{(n)}, \delta_n, \rho)],$$

where $\varphi(z)$ is the function constructed in Lemma 2. Since the polydiscs $P(z^{(n)}, \frac{1}{8}, \frac{1}{8})$ do not intersect, the function $\Phi(z)$ is plurisubharmonic in $\mathbf{C}^2$. It is clear that a) and b) of the lemma are satisfied. Further,

$$m\left(\bigcup_{n=1}^{N} P\left(z^{(n)}, \tfrac{1}{8}, \delta_n\right)\right) = \sum_{n=1}^{N} 2^{-6}\pi^2\delta_n^2 r_n^4. \tag{2}$$

Since $\delta_\lambda \to 0$ as $n \to \infty$ and $r_{n+1} \ge 2r_n$ for all n, (2) is $o(r_N^4)$. This proves c). Finally, for each point $\zeta \in B(\zeta^{(n)}, \tau/2)$, points of the form $t\zeta$, where $|t - r_n| \le r_n/15$, fall in the polydisc $P(z^{(n)}, 1/15 + 8\tau_n/15; \tau_n/2)$. This together with the inequality $1/r_n \le \tau_n/2 \le 1/124$, which is satisfied by the construction of the sequences r_n and τ_n, implies that $B(t\zeta, 1) \subset P(z^{(n)}; \frac{1}{12}, \tau_n)$. It remains to note that any point $\zeta \in S \cap \Gamma$ belongs to an infinite number of the balls $B(\zeta^{(n)}; \tau_n/2)$. This proves the lemma.

PROOF OF THE THEOREM. Let $f(w)$ be an entire function in the plane with order ρ and indicator identically equal to 1.([1]) Then uniformly for $\varphi \in [0, 2\pi)$ we have

$$\lim_{\substack{t\to\infty \\ te^{i\varphi}\notin E}} t^{-\rho} \ln|f(te^{i\varphi})| = \overline{\lim_{t\to\infty}}\, t^{-\rho} \ln|f(te^{i\varphi})| = 1,$$

where E is a set on the plane which can be covered by a sequence of discs $B(w_j, r_j)$ such that

$$\sum_{f:|w_j|<R} r_j = o(R).$$

([1])For the existence of such functions and their properties, see [1].

Therefore we can construct a continuous function $\alpha(t) \geq 0$ with $\alpha(t) \to 0$ as $t \to \infty$ such that

$$1 - \alpha(t) \leq t^{-\zeta} \ln |f(te^{i\varphi})| \leq 1 + \alpha(t), \tag{3}$$

where the right inequality is satisfied for all t and φ, and the left for $te^{i\varphi} \notin E$. We set

$$\beta(t) = \max\{\alpha(t-1) + (t-1)^{-1}\rho; \alpha(t) + 3t^{-\rho}\ln(1+3t)\}$$

for $t \geq 2$, and $\beta(t) = \beta(2)$ for $t < 2$. We select a monotonic twice continuously differentiable function $\gamma(t) \geq \beta(t)$ such that[2]

$$\gamma(t) \to 0 \quad \text{as } t \to \infty, \tag{4}$$

$$t\gamma'(t) \to 0 \quad \text{as } t \to \infty, \tag{5}$$

$$t^2\gamma''(t) \to 0 \quad \text{as } t \to \infty. \tag{6}$$

It follows from (4)–(6) that there is $t' > 0$ so that the function $|w|^\rho(1 - 2\gamma(|w|))$ is subharmonic for $|w| \geq t'$. This function is subharmonic for all $w \in \mathbf{C}$ if we redefine the function $\gamma(t)$ for $t < t'$ by setting $\gamma(t) = \frac{1}{2} - \frac{1}{2}[1 - 2\gamma(t')t'^\rho]t^{-\rho}$. Further, for $t \geq t'$ we have $[1 - \gamma(t+1)](t+1)^\rho \leq [1 - \alpha(t)]t^\rho$. Therefore (3) implies that outside the set $E' = E \cup \{w : |w| < t'\}$ we have

$$|f(w)| \geq \sup_{|\lambda|<1} \exp\{[1 - \gamma(|w+\lambda|)]|w+\lambda|^\rho\}, \tag{7}$$

and for all $|w| \geq t'$,

$$|f(w)| \leq \exp\{[1 + \gamma(|w|)]|w|^\rho\}(1 + 3|w|)^{-3}. \tag{8}$$

Let $\chi(w)$ be a function in $C^\infty(\mathbf{C})$ such that $\chi(w) = 1$ for $|w| \leq 1$, $\chi(w) = 0$ for $|w| \geq 3/2$, and $0 \leq \chi(w) \leq 1$ for all w. In Lemma 3 we choose as $\Delta(t)$ any function monotonically decreasing to zero such that $\Delta(t) \geq \max\{\gamma(2t); -t\gamma'(2t)\}^{1/3}$ for all t. Using the notation of Lemma 3, we set

$$\kappa(z) = 2\sum_{n=n'}^{\infty} \chi(6\lambda(z; z^{(n)}))\chi(2^{-1}\delta_n^{-1}\mu(z; z^{(n)})),$$

$$\Psi(z) = \tfrac{1}{2}\Phi(z) + \frac{1}{2}(|z_1|^\rho + |z_2|^\rho) + [\kappa(z) - 1][\gamma(|z_1|)|z_1|^\rho + \gamma(|z_2|)|z_2|^\rho],$$

where the functions $\lambda(z; z')$ and $\mu(z; z')$ are defined by (1), and $n' = n'(\rho)$ is some fixed constant to be chosen below.

Since the function $|w|^\rho[\frac{1}{2} - \gamma(|w|)]$ is subharmonic in $\mathbf{C}$, the function $\Psi(z)$ is automatically plurisubharmonic for $z \notin \operatorname{supp}\kappa(z)$. Further, for $i, j = 1, 2$

---

[2] For example, we may take

$$\gamma(t) = \frac{1}{t}\int_0^t \frac{1}{s}\int_0^s \max_{r \geq q}\{\beta(r)\}\, dq\, ds.$$

and $z \in P(z^{(n)}, \frac{1}{4}, \frac{1}{4})$ we have as $n \to \infty$ that

$$\frac{\partial \kappa(z)}{\partial z_j} = O(\delta_n^{-1}|z^{(n)}|^{-1}) = O(\Delta^{-1}(|z^{(n)}|)|z^{(n)}|^{-1};$$
$$\frac{\partial^2 \kappa(z)}{\partial z_i \partial z_j} = O(\delta_n^{-2}|z^{(n)}|^{-2}) = O(\Delta^{-2}(|z^{(n)}|) \times |z^{(n)}|^{-2}).$$

It now follows from these relations, from the definition of the function $\Delta(t)$, and from (4)–(6) that for $z \in P(z^{(n)}, \frac{1}{4}, \frac{1}{4})$ the Levi form of the function

$$[\kappa(z) - 1][\gamma(|z_1|)|z_1|^\rho + \gamma(|z_2|)|z_2|^\rho]$$

is $o(|z^{(n)}|^{\rho-2})$ as $n \to \infty$. Therefore, by Lemma 1, we can choose n' so that $\Psi(z)$ is plurisubharmonic in $\mathbf{C}^2$.

We now set

$$A_2 = \bigcup_{n=n'}^{\infty} P(z^{(n)}; \frac{1}{6}; 2\Delta_n);$$
$$A_3 = \bigcup_{n=n'}^{\infty} P\left(z^{(n)}; \frac{1}{4} + \frac{1}{|z^{(n)}|}; 3\Delta_n + \frac{1}{|z^{(n)}|}\right).$$

It is not difficult to show (see the proof of c) of Lemma 3) that the set A_3 has zero relative measure. It follows from (7) and (8) that outside the set $A_4 \stackrel{\text{def}}{=} A_1 \cup \{z : z_1 \in E'\} \cup \{z; z_2 \in E'\}$, which has also zero relative measure, we have

$$\sup_{|z-\zeta|<1} \exp\Psi(\zeta) \le |f(z_1)| \cdot |f(z_2)|, \tag{9}$$

and in the set $A_2 \backslash A_1$, where A_1 is the set from Lemma 3, we have

$$\exp\Psi(z) \ge |f(z_1)f(z_2)| \cdot (1 + |z|)^6. \tag{10}$$

We construct a function $\eta(z) \in C^\infty(\mathbf{C}^2)$ such that $0 \le \eta(z) \le 1$ for all z, $\eta(z) = 0$ for $z \in A_1$, $\eta(z) = 1$ for $z \notin A_2$, and

$$\|\overline{\partial}\eta\| = \left(\left|\frac{\partial\eta}{\partial\overline{z}_1}\right|^2 + \left|\frac{\partial\eta}{\partial\overline{z}_2}\right|^2\right)^{1/2} \le 1 \quad \text{for all } z \in \mathbf{C}^2.$$

We set $F(z) = \eta(z)f(z_1)f(z_2) - V(z)$, where $V(z)$ is a solution to the $\overline{\partial}$-problem

$$\overline{\partial}V(z) = f(z_1)f(z_2)\overline{\partial}\eta(z) \tag{11}$$

in the space $L^2(\mathbf{C}^2)$ with weight $\exp\{-2\Psi(z)\}$. This solution exists, since by (10) we have

$$\int_{C^2} \|\overline{\partial}\eta(z)\| \cdot |f(z_1)f(z_2)|^2 \exp\{-2\Psi(z)\}(1 + |z|^2)^2\, dm$$
$$\le \int_{A_2\backslash A_1} |f(z_1)f(z_2)|^2 \exp\{-2\Psi(z)\}(1 + |z|)^4\, dm \le \pi^2/6.$$

By a theorem of Hörmander ([12], Theorem 4.4.3) there exists a function $V(z)$ satisfying (11) for which

$$\int_{C^2} |V(z)|^2 \exp\{-2\Psi(z)\}\, dm \le \pi^2/12. \tag{12}$$

Let the point z be chosen so that $B(z,1) \cap \operatorname{supp} \overline{\partial}\eta = \phi$. Then the function $V(z)$ is analytic in the ball $B(z,1)$, and consequently

$$\begin{aligned} |V(z)| &\le \frac{2}{\pi^2} \int_{|z-\zeta|<1} |V(\zeta)|\, dm \\ &\le \frac{2}{\pi^2} \left[\int_{|z-\zeta|<1} \exp 2\Psi(\zeta)\, dm \right]^{1/2} \left\{ \int_{|z-\zeta|<1} |V(\zeta)|^2 \exp(-2\Psi(\zeta)) dm \right\}^{1/2}. \end{aligned}$$

By (12) the integral in the braces is no greater than $\pi^2/12$; therefore

$$|V(z)| \le \frac{1}{\pi\sqrt{3}} \sup_{|z-\zeta|<1} \exp \Psi(\zeta). \tag{13}$$

This together with (9) implies that for $z \notin V_4$ we have

$$|F(z) - f(z_1)f(z_2)| = |V(z)| \le \frac{1}{\pi\sqrt{3}} |f(z_1)f(z_2)|. \tag{14}$$

We now note that for all $z \in \mathbf{C}^2$

$$\begin{aligned} F(z) &\le \frac{2}{\pi^2} \int_{|\zeta-z|<1} |V(\zeta) + \eta(\zeta) f(\zeta_1) f(\zeta_2)|\, dm \\ &\le \frac{2}{\pi^2} \int_{|\zeta-z|<1} |V(\zeta)|\, dm + \frac{2}{\pi^2} \int_{|\zeta-z|<1} |f(\zeta_1)f(\zeta_2)|\, dm \\ &= I_1 + I_2. \end{aligned}$$

As above, the quantity I_1 for all $z \in \mathbf{C}^2$ is bounded from above by

$$\sup_{|z-\zeta|<1} \exp \Psi(\zeta),$$

which means that it does not grow faster than $K_3 \exp\{3|z|^\rho\}$. The quantity I_2 admits a similar bound. Consequently the function $F(z)$ has growth no greater than normal type relative to the order ρ. We note that for any entire function $g(z)$ of normal type relative to order ρ and for any set D with zero relative measure we have([3])

$$\int_{\{z \in B(1): tz \in D\}} \ln |g(tz)|\, dm = o(t^\rho).$$

Applying this statement to the functions $F(z)$ and $f(z_1)f(z_2)$, and using (14) we conclude that the function $F(z)$ has completely regular growth in sense (A), where for almost all $z \in \mathbf{C}^2$

$$L_r(z; F) = L_r(z; f(z_1)f(z_2)) = |z_1|^\rho + |z_2|^\rho. \tag{15}$$

([3])Equivalent statements are proved, for example in [3], p. 9, and [7], §4.

Further, it follows from the definition of the function $\Psi(z)$ that, like $\Phi(z)$, it satisfies condition d) of Lemma 3 (with ε replaced by $\varepsilon/3$). Since $\operatorname{supp}\bar\partial\eta\cap A_1 = \phi$, we obtain using (13) that for all $\zeta \in S\cap\Gamma$ there is a sequence $t' \to \infty$ such that for $t \in [t' - t'/15, t' + t'/15]$ we have

$$|F(t\zeta)| = |V(t\zeta)| \leq \exp\{(1 - \varepsilon/3)[(t|\zeta_1| + 1)^\rho + (t|\zeta_2| + 1)^\rho]\}.$$

Consequently, for those $z \in \Gamma$ for which (15) holds, i.e., for almost all $z \in \Gamma$, the function $F(z \cdot w)$ is not a function of completely regular growth relative to the variable $w \in \mathbf{C}$.

Finally, let T be the automorphism of $\mathbf{C}^2$ given by

$$(z_1, z_2) \to \left(\frac{z_1 + z_2}{\sqrt{2}}, \frac{z_1 - z_2}{\sqrt{2}}\right).$$

The function $\tilde{F}(z) = F(z)F(Tz)$ has order ρ and has completely regular growth in sense (A), but its restriction to the rays $\{z \cdot w : w \in \mathbf{C}\}$ for almost all $z \in \Gamma \cup T^{-1}(\Gamma)$ is not a function of completely regular growth relative to w. It remains to note that $\Gamma \cup T^{-1}(\Gamma) = \mathbf{C}^2\backslash\{0\}$. This proves the theorem.

References

1. B. Ya. Levin, *Distribution of zeros of entire functions*, GITTL, Moscow, 1956; English transl., Amer. Math. Soc., Providence, R.I., 1964.

2. V. S. Azarin, *Subharmonic functions of completely regular growth in multidimensional space*, Dokl. Akad. Nauk SSSR **146** (1962), 743–746; English transl. in Soviet Math. Dokl. **3** (1962).

3. P. Z. Agranovich and L. I. Ronkin, *Functions of completely regular growth of several variables*, Preprint, Fiz.-Tekhn. Inst. Nizkikh Temperatur Akad. Nauk Ukrain. SSR, Kharkov, 1976. (Russian) R. Zh. Mat. **1977**, 56112.*

4. V. S. Azarin, *Asymptotic behavior of subharmonic and entire functions*, Dokl. Akad. Nauk SSSR **229** (1976), 1289–1291; English transl. in Soviet Math. Dokl. **17** (1976).

5. P. Z. Agranovich, *On a criterion for regularity of growth*, Preprint No. 30–77, Fiz.-Tekhn. Inst. Nizkikh Temperatur Akad. Nauk Ukrain. SSR, Kharkov, 1977. (Russian) R. Zh. Mat. **1978**, 76230.

6. ——, *Functions of several variables with completely regular growth*, Teor. Funktsiĭ, Funktsional. Anal. i Prilozhen. Vyp. **30** (1978), 3–13. (Russian)

7. S. Yu. Favorov, *On the addition of the indicators of entire and subharmonic functions of several variables*, Mat. Sb. **105(147)** (1978), 128–140; English transl. in Math. USSR Sb. **34** (1978).

8. Lawrence Gruman, *Entire functions of several variables and their asymptotic growth*, Ark. Mat. **9** (1971), 141–163.

9. B. O. Berndtsson, *On the asymptotic growth of (pluri)subharmonic functions*, Preprint No. 1976-10, Göteborgs Univ., Göteborg, 1976.

10. V. V. Morzhakov, *Differential operators of infinite order in spaces of holomorphic functions of several complex variables*, Author's summary of Candidate's dissertation, Rostov-on-Don, 1974. (Russian)

11. S. Yu. Favorov, *Reduction sets for subharmonic functions of completely regular growth*, Sibirsk. Mat. Zh. **20** (1979), 1294–1302; English transl. in Siberian Math. J. **20** (1979).

12. Lars Hörmander, *An introduction to complex analysis in several variables*, Van Nostrand, Princeton, N.J., 1966.

Translated by J. NUNEMACHER

* *Editor's note.* Compare their paper of the same title, Ann. Polon. Math. **39** (1981), 239–254.

Amer. Math. Soc. Transl.
(2) Vol. **144**, 1989

On the Theory of the One-Point Boundary Value Problem for the Laplace Operator. I

UDC 513.88

V. È. LYANTSE AND KH. B. MAĬORGA

0. Let $D = C_0^\infty(\mathbf{R}^3)$, D' be the dual space to D of distributions, and Δ be the Laplace operator $D' \to D'$. Thus, for all $f \in D'$ and all $\varphi \in D$, $(\Delta f|\varphi) = (f|\Delta\varphi)$, where $(f|\varphi)$ denotes the value of the functional f on the test function $\bar\varphi$, where $\bar\varphi(x) \overset{\text{def}}{=} \overline{\varphi(x)}$ for all $x \in \mathbf{R}^3$.

This definition of Δf is equivalent to the following. Let $\hat f \overset{\text{def}}{=} \Phi f$ be the Fourier image of the distribution f, where, for all $\varphi \in D$,

$$\hat\varphi(\xi) = \Phi\varphi(\xi) \overset{\text{def}}{=} \frac{1}{(2\pi)^{3/2}} \int_{\mathbf{R}^3} \varphi(x) e^{-ix\xi}\, dx, \qquad \xi \in \mathbf{R}^3.$$

Then $\widehat{\Delta f}(\xi) = -\xi^2 \hat f(\xi)$, i.e., $\widehat{\Delta f}$ is the result of multiplying $\hat f$ by the multiplier $-\xi^2$, where $\xi^2 = \xi_1^2 + \xi_2^2 + \xi_3^2$.

We denote by H^s the Sobolev space of order s: for all $s \in \mathbf{R}$,

$$H^s = H^s(\mathbf{R}^3) \overset{\text{def}}{=} \left\{ f \in L_2(\mathbf{R}^3) \colon \int_{\mathbf{R}^3} (1+\xi^2)^s |\hat f(\xi)|^2\, d\xi < \infty \right\};$$

for all $f, g \in H^s$,

$$(f|g) \overset{\text{def}}{=} \int_{\mathbf{R}^3} (1+\xi^2)^s \hat f(\xi) \overline{\hat g(\xi)}\, d\xi. \tag{0.1}$$

Let S be the natural restriction of the operator $-\Delta$ to $H \overset{\text{def}}{=} H^0 = L_2(\mathbf{R}^3)$, so that $D(S) \overset{\text{def}}{=} \{f \in H \colon \Delta f \in H\}$ and $Sf = -\Delta f$ for $f \in D(S)$. It is clear that $D(S) = H^2$, that S is a selfadjoint operator $H \to H$, and that the inner product([1]) $(\cdot|\cdot)_{D[S]}$ is equivalent to $(\cdot|\cdot)_{H^2}$.

1980 *Mathematics Subject Classification* (1985 *Revision*). Primary 35J05, 35J25; Secondary 47F05.

Translation of Teor. Funktsiĭ, Funktsional. Anal. i Prilozhen. Vyp. 38(1982), 84–91; MR **84d**: 35041.

([1])Let L be a linear operator $H \to H$. We denote by $D[L]$ the domain $D(L)$ of the operator L, endowed with the inner product of its graph: $(f|g)_{D[L]} = (f|g) + (Lf|Lg)$, if $f, g \in D(L)$.

From elementary theorems on the degree of smoothness and degree of decay of the Fourier transform, it is easy to derive the following assertion.

0.1. LEMMA. *If $f \in H^2$, then f is continuous on $\mathbf{R}^3$ and $f(x) \to 0$ as $x \to \infty$. The Dirac δ distribution (for all $\varphi \in D$, $\delta(\varphi) = \varphi(0)$) is continuous with respect to $\|\cdot\|_{H^2}$ (and therefore with respect to $\|\cdot\|_{D[S]}$ as well).*

Note that the analogous lemma does not hold for $\mathbf{R}^n$ with $n > 3$.

Let $D(S_0) \stackrel{\text{def}}{=} \{f \in D(S): f(0) = 0\}$ and $S_0 \subset S$. From Lemma 0.1 it follows that S_0 is a densely defined, closed, symmetric operator $H \to H$ with deficiency index (1.1). In [1], it is explained how the selfadjoint extensions of S_0 are connected with particle interaction theory.(2) Here, we supplement the description of selfadjoint extensions of S_0 given in [1], describing these extensions in terms of the boundary conditions, both in the "momentum" ξ-representation and in the "position" x-representation.(3) In addition, for each such selfadjoint (and even quasi-selfadjoint) extension, we explicitly construct the operator diagonalizing this extension, i.e., we construct the corresponding inversion formulas(4) (and the corresponding spectral density).

1. First, let us describe the operator $S_1 \stackrel{\text{def}}{=} S_0^*$. Note that $S_0 \subset S \subset S_1$.

1.1. THEOREM. *Let*

$$\pi_0 \stackrel{\text{def}}{=} (4\pi\sqrt{2})^{1/2}, \tag{1.0}$$

$$e_0(x) \stackrel{\text{def}}{=} \frac{\pi_0}{4\pi|x|} \sin\frac{|x|}{\sqrt{2}} \exp\frac{-|x|}{\sqrt{2}}, \qquad x \in \mathbf{R}^3. \tag{1.1}$$

Then

$$e_0 \in H^2 = D(S); \qquad \|e_0\|_{D[S]} = 1 \quad \text{and} \quad D(S) = D(S_0) \oplus Ce_0. \tag{1.2}$$

Moreover,

$$Se_0(x) = \frac{\pi_0}{4\pi|x|} \cos\frac{|x|}{\sqrt{2}} \exp\frac{-|x|}{\sqrt{2}}, \qquad x \in \mathbf{R}^3, \tag{1.3}$$

$$Se_0 \in D(S_1); \qquad \|Se_0\|_{D[S_1]} = 1 \quad \text{and} \quad D(S_1) = D(S) \oplus CSe_0. \tag{1.4}$$

The direct sum decompositions (1.2) *and* (1.4) *are orthogonal with respect to* $(\cdot|\cdot)_{D[S_1]}$. *Finally,*

$$S_1 Se_0 = -e_0. \tag{1.5}$$

For the proof of Theorem 1.1 we use an observation made in [11]. Let H be an abstract Hilbert space, let $C(H)$ be the class of linear, densely defined, closed operators $H \to H$, let $L_0, L \in C(H)$ with $L_0 \subset L$, and let $M \stackrel{\text{def}}{=} L_0^*$ and $M_0 \stackrel{\text{def}}{=} L^*$, so that $M_0, M \in C(H)$ and $M_0 \subset M$. Consider the direct sum decompositions $D(L) = D(L_0) \oplus U$ and $D(M) = D(M_0) \oplus V$, which are orthogonal with respect to $(\cdot|\cdot)_{D[L]}$ and $(\cdot|\cdot)_{D[M]}$, respectively.

(2)See also [2]–[5]; in the recent paper [6] a description of selfadjoint extensions of S_0 is obtained by the methods of nonstandard analysis.

(3)Note that the description of the boundary conditions in the x-representation given by us below is contained implicitly in [7] and [8], as well as explicitly in [9] and [10].

(4)These inversion formulas can be viewed as explicit inversion formulas for the "perturbed" (by a special procedure) Fourier transform.

1.2. ASSERTION. *The restriction of L to U is a unitary transformation of U onto V relative to $(\cdot|\cdot)_{D[L]}$ and $(\cdot|\cdot)_{D[M]}$. Its inverse is the restriction of $-M$ to V. In particular, $D(L) = D(L_0) \oplus MV$ and, for all $v \in V$, $LMv = -v$.*

PROOF OF THEOREM 1.1. We apply Assertion 1.2 to $L = S_1$ and $L_0 = S$. In this case $M = S$, $M_0 = S_0$, and $V = D(S) \ominus D(S_0)$. However,([5]) $D(S_0) = Z(\delta)$, where δ is the Dirac functional on $D(S) = H^2$. Therefore, $V = R(\delta^*)$, where $\delta^*: C \to D(S)$. Thus, $D(S_1) = D(S) \oplus SR(\delta^*)$, where δ^* satisfies the identity $(\delta^* c|f)_{D[S]} = c\overline{f(0)}$ for all $f \in D(S)$ and all $c \in C$. Putting $e \overset{\text{def}}{=} \delta^* 1$, we arrive at the $(\cdot|\cdot)_{D[S_1]}$-orthogonal decomposition $D(S_1) = D(S) \oplus CSe$ and the relation

$$\forall f \in H^2 \quad (f|e)_{D[S]} = f(0). \tag{1.6}$$

Taking into account that $f(0) = (2\pi)^{-3/2} \int_{\mathbf{R}^3} \hat{f}(\xi)\, d\xi$, from (1.6) we get the following formula for the Fourier images:

$$\hat{e}(\xi) = \frac{1}{(2\pi)^{3/2}} \frac{1}{1+\xi^4}; \qquad S\hat{e}(\xi) = \frac{1}{(2\pi)^{3/2}} \frac{\xi^2}{1+\xi^4}. \tag{1.7}$$

On carrying out the inverse Fourier transform, we find that $e = \pi_0^{-1} e_0$ and $Se = \pi_0^{-1} Se_0$, where π_0, e_0, and Se_0 are defined by means of (1.0), (1.1), and (1.3). Indeed, let $\hat{E}_\xi(\xi) \overset{\text{def}}{=} -(2\pi)^{-3/2}(\xi^2 - \zeta)^{-1}$, so that $(\Delta + \zeta)E_\zeta = \delta$. Then (see, for example, [12]),

$$E_{k^2}(x) = -(4\pi|x|)^{-1} \exp(\pm ik|x|),$$

where the sign in front of i should be chosen so that, for $k \in C$ given, $E_{k^2}^2 \in H$. In particular,

$$E_{\pm i}(x) = -(4\pi|x|)^{-1} \exp[-(1 \mp i)|x|/\sqrt{2}].$$

Our formulas for e_0 and Se_0 follow from the fact that $\hat{e} = (2i)^{-1}[\hat{E}_{-i} - \hat{E}_{+i}]$ and $S\hat{e} = -2^{-1}[\hat{E}_{-i} + \hat{E}_{+i}]$.

From the equality $LMv = -v$ mentioned in Assertion 1.2 we get (1.5). To derive the orthogonal decomposition (1.2), Assertion 1.2 has to be applied to the pair of operators S_0, S.

Let us substitute e for f in (1.6). We obtain $\|e\|_{D[S]}^2 = e(0)$, and since $e_0 = \pi_0 e$, we have $\|e_0\|_{D[S]}^2 = \pi_0 e_0(0)$. But from (1.1) it is seen that $e_0(0) = \pi_0^{-1}$, and thus $\|e_0\|_{D[S]} = 1$. Hence, by (1.5) we get $\|Se_0\|_{D[S_1]} = 1$.

1.3. COROLLARY. *On $D(S_1)$ there are linear functionals μ and ν, continuous with respect to $\|\cdot\|_{D[S_1]}$, such that*

$$\begin{aligned} &\forall f \in D(S_1) \quad f - \mu(f)Se_0 \in D(S); \\ &f - \mu(f)Se_0 - \nu(f)e_0 \in D(S_0). \end{aligned} \tag{1.8}$$

Namely,

$$\mu(f) = (f|Se_0)_{D[S_1]}; \qquad \nu(f) = (f|e_0)_{D[S_1]}. \tag{1.8'}$$

([5])We denote the kernel of an arbitrary linear operator T by $Z(T)$, and its range by $R(T)$.

In particular,

$$\nu(Se_0) = \nu(e_0) = 1; \qquad \mu(e_0) = \nu(Se_0) = 0. \tag{1.8''}$$

The description of the operator S_1 can be summed up as follows.

1.4. REMARK. Let $\hat{S}_1$ be the Fourier image of the operator S_1, i.e., $D(\hat{S}_1) \stackrel{\text{def}}{=} \Phi D(S_1)$, and $\hat{S}_1 \hat{f} \stackrel{\text{def}}{=} \widehat{S_1 f}$ for $f \in D(S_1)$. Then $D(\hat{S}_1)$ consists precisely of those $f \in H$ for which there exists $k \in C$ such that

$$\int_{\mathbf{R}^3} |\xi^2 \hat{f}(\xi) - k|^2 \, d\xi < \infty. \tag{1.9}$$

If k is one such number, then

$$\hat{S}_1 \hat{f}(\xi) = \xi^2 \hat{f}(\xi) - k \tag{1.10}$$

and

$$k = \frac{\pi_0}{(2\pi)^{3/2}} \mu(f), \tag{1.10'}$$

where μ is the functional defined by condition (1.8).

PROOF. Let (1.9) be satisfied for some $k \in C$. Then, defining $\mu(f)$ by (1.10′), we find that $f - \mu(f) Se_0 \in D(S)$, and thus $f \in D(S_1)$. Conversely, let $f \in D(S_1)$. Then, by (1.5),

$$S_1 f = S(f - \mu(f) Se_0) - \mu(f) e_0$$

and

$$S_1 \hat{f}(\xi) = \xi^2 [\hat{f}(\xi) - \mu(f) S\hat{e}_0(\xi)] - \mu(f) \hat{e}_0(\xi) = \xi^2 \hat{f}(\xi) - \frac{\pi_0}{(2\pi)^{3/2}} \mu(f).$$

1.5. COROLLARY. *$D(S_1)$ consists precisely of those $f \in H$ for which there exists $\lambda \in \mathbf{C}$ such that $-\Delta f - \lambda \delta \in H$, where Δ is the Laplace operator $D' \to D'$, and δ is the Dirac distribution. If λ is such a number, then*

$$S_1 f = -\Delta f - \lambda \delta \tag{1.11}$$

and

$$\lambda = \pi_0 \mu(f). \tag{1.11'}$$

1.6. REMARK. For all $f \in D(S_1)$, the numbers $\mu(f)$ and $\nu(f)$ are uniquely determined by the conditions

$$\int_{\mathbf{R}^3} (1 + \xi^2)^2 \left| \hat{f}(\xi) - \frac{\pi_0}{(2\pi)^{3/2}} \frac{\mu(f) \xi^2}{\xi^4 + 1} \right|^2 d\xi < \infty, \tag{1.12}$$

$$\int_{\mathbf{R}^3} \left[\hat{f}(\xi) - \frac{\pi_0}{(2\pi)^{3/2}} \frac{\mu(f) \xi^2 + \nu(f)}{\xi^4 + 1} \right] d\xi = 0. \tag{1.13}$$

2. It turns out that the functionals μ and ν have the following x-representation.

2.1. THEOREM. *For all $f \in D(S_1)$ the limits*

$$\mu_0(f) \stackrel{\text{def}}{=} \lim_{x\to 0} |x| f(x), \tag{2.1}$$

$$\nu_0(f) \stackrel{\text{def}}{=} \lim_{x\to 0} \left[f(x) - \frac{\mu_0(f)}{|x|} \right] \tag{2.2}$$

exist and are finite. The functionals μ and ν are expressed in terms of μ_0 and ν_0 as follows (see (1.0)*):*

$$\mu(f) = \frac{4\pi}{\pi_0} \mu_0(f), \tag{2.3}$$

$$\nu(f) = \pi_0 \left[\nu_0(f) + \frac{1}{\sqrt{2}} \mu_0(f) \right]. \tag{2.4}$$

In particular, for all $f \in D(S) = H^2$

$$\mu_0(f) = 0, \qquad \nu_0(f) = \delta(f) = f(0). \tag{2.5}$$

PROOF. Let $f \in D(S_1)$, so that $f - \mu(f)Se_0 \in D(S)$. According to Lemma 0.1, the function $f - \mu(f)Se_0$ is continuous, and thus $|x|[f(x) - \mu(f)Se_0(x)] \to 0$ as $x \to 0$. But, by (1.3), $|x|Se_0(x) \to \pi_0/(4\pi)$ as $x \to 0$. Consequently the limit (2.1) exists, and (2.3) holds. Further, taking into account that $e_0 = \pi_0 e$ and that $(Se_0|e_0)_{D[S_1]} = 0$, from (1.6) and (1.8″) we obtain that for all $f \in D(S_1)$,

$$\nu(f) = (f - \mu(f)Se_0|e_0)_{D[S_1]} = \pi_0 \delta(f - \mu(f)Se_0).$$

However, from (1.3) it follows that

$$\lim_{x\to 0} [\tfrac{\pi_0}{4\pi|x|} - Se_0(x)] = \pi_0^{-1}.$$

Therefore, the limit (2.2) exists and (2.4) holds.

2.2. REMARK. Let $f \in D(S_1)$. Then

$$\hat{S}_1 \hat{f}(\xi) = \xi^2 \hat{f}(\xi) - \sqrt{\tfrac{2}{\pi}} \mu_0(f), \qquad S_1 f = -\Delta f - 4\pi \mu_0(f)\delta. \tag{2.6}$$

This follows immediately from 1.4–1.6 and 2.1.

3. The numbers $\mu_0(f)$ and $\nu_0(f)$ (as well as $\mu(f)$ and $\nu(f)$) are naturally viewed as boundary values (the trace at the point $x = 0$ of the function $f \in D(S_1)$). It turns out that the following Green's formula holds.

3.1. ASSERTION. *For all $f, g \in D(S_1)$,*

$$(S_1 f|g) - (f|S_1 g) = \nu(f)\overline{\mu(g)} - \mu(f)\bar{\nu} g = 4\pi[\nu_0(f)\overline{\mu_0(g)} - \mu_0(f)\overline{\nu_0(g)}]. \tag{3.1}$$

PROOF. Let $f, g \in D(S_1)$, so that $g - \mu(g)Se_0 - \nu(g)e_0 \in D(S_0)$. Since $S_0 = S_1^*$, we have

$$\begin{aligned}(S_1 f|g) &= (f|S_0[g - \mu(g)Se_0 - \nu(g)e_0]) + (S_1 f|Se_0)\overline{\mu(g)} + (S_1 f|e_0)\overline{\nu(g)} \\ &= (f|S_1 g) + (f|e_0)_{D[S_1]}\overline{\mu(g)} - (f|Se_0)_{D[S_1]}\overline{\nu(g)}.\end{aligned}$$

Taking (1.8′) into account, we arrive at (3.1).

3.2. Definition. For all $\theta \in \mathbf{C}$, let

$$D(S^\theta) \stackrel{\text{def}}{=} \{f \in D(S_1) : \nu_0(f) = \theta\mu_0(f)\},$$

and $S^\theta f = S_1 f$ for $f \in D(S^\theta)$. In addition, put $S^\infty \stackrel{\text{def}}{=} S$.

From Green's formula (3.1), the next assertion follows in the obvious fashion.

3.3. Assertion. *The set of all selfadjoint extensions of the operator S_0 coincides with the set of all S^θ such that $\theta \in \mathbf{R}$ or $\theta = \infty$.*

4. Let us describe the spectrum and resolvent of the operator S^θ. For brevity, we denote the operator S^θ by T. Then, putting, for all $f \in D(S_1)$,

$$\gamma_0(f) = \nu_0(f) - \theta\mu_0(f), \tag{4.1}$$

we have (see 3.2)

$$D(T) \stackrel{\text{def}}{=} \{f \in D(S_1) \colon \gamma_0(f) = 0\}, \qquad T \subset S_1. \tag{4.2}$$

We are not assuming that $\theta \in \mathbf{R}$, so T need not be selfadjoint.

4.1. Lemma. *The number ζ is an eigenvalue of the operator S_1 if and only if $\zeta \in \mathbf{C}\backslash[0,\infty[$. Each eigenvalue is simple, and the corresponding eigenfunction E_ζ is given by*

$$E_\zeta(x) = -\frac{1}{4\pi|x|}e^{i\sqrt{\zeta}|x|}, \qquad x \in \mathbf{R}^3. \tag{4.3}$$

Henceforth, the value of $\sqrt{\zeta}$ (for $\zeta \in \mathbf{C}\backslash[0,\infty[$) is chosen so that $\operatorname{Im}\sqrt{\zeta} > 0$.

Proof. According to (2.6) the equation $S_1 u = \zeta u$ is equivalent to the equation $\xi^2\hat{u}(\xi) - \sqrt{2/\pi}\,\mu_0(u) = \zeta\hat{u}(\xi)$, whence $\hat{u}(\xi) = \text{const}\,\hat{E}_\zeta(\xi)$, where

$$\hat{E}_\zeta(\xi) = -\frac{1}{(2\pi)^{3/2}}\cdot\frac{1}{\xi^2 - \zeta}, \qquad \xi \in \mathbf{R}^3. \tag{4.3′}$$

Clearly, $\hat{E}_\zeta \in H$ if and only if $\zeta \in \mathbf{C}\backslash[0,\infty[$. Carrying out the inverse Fourier transform (see the proof of Theorem 1.1), from (4.3′) we get (4.3).

4.2. Lemma. *For arbitrary ζ, $\zeta_1 \in \mathbf{C}\backslash[0,\infty[$,*

$$(E_\zeta|E_{\zeta_1}) = \frac{1}{4\pi(\sqrt{\zeta} + \sqrt{\bar{\zeta}_1})}. \tag{4.4}$$

Proof. By 4.1,

$$(E_\zeta|E_{\zeta_1}) = \zeta^{-1}(S_1E_\zeta|E_{\zeta_1}).$$

Next, we apply Green's formula (3.1), taking into account the fact that, by (2.1), (2.2), and (4.3),

$$\mu_0(E_\zeta) = -\frac{1}{4\pi}, \qquad \nu_0(E_\zeta) = -\frac{i\sqrt{\zeta}}{4\pi}. \tag{4.5}$$

Moreover, it should be noted that $\overline{\sqrt{\bar\zeta}} = -\sqrt{\zeta}$, since we stipulated that $\sqrt{\zeta}$ be chosen in the upper half plane.

4.3. DEFINITION. For all $\zeta \in \mathbf{C}\backslash[0,\infty[$,

$$\gamma(\zeta) \stackrel{\text{def}}{=} \gamma_0(E_\zeta) = \nu_0(E_\zeta) - \theta\mu_0(E_\zeta)$$

(see (4.1)). By (4.5) we have

$$\gamma(\zeta) = \frac{(i\theta + \sqrt{\zeta})}{4\pi i}. \tag{4.5'}$$

4.4. LEMMA. *Let $u \in D(S_1)$, $\zeta \in \mathbf{C}\backslash[0,\infty[$, and $(S_1 - \zeta)u = f$. Then*

$$\hat{u}(\xi) = \frac{1}{\xi^2 - \zeta}\left[\hat{f}(\xi) - \sqrt{\frac{2}{\pi}}\mu_0(u)\right]; \tag{4.6}$$

$$\gamma_0(u) = -(f|E_{\bar\zeta}) - 4\pi\gamma(\zeta)\mu_0(u). \tag{4.6'}$$

PROOF. (4.6) follows from (2.6). Put $\hat{u}_1(\xi) \stackrel{\text{def}}{=} (\xi^2 - \zeta)^{-1}\hat{f}(\xi)$. Then $u_1 \in D(S)$, and therefore (see (2.5))

$$\nu_0(u_1) = \delta(u_1) = (2\pi)^{3/2}\int_{\mathbf{R}^3}\hat{u}_1(\xi)\,d\xi = -(f|E_{\bar\zeta}).$$

Putting $u_2 \stackrel{\text{def}}{=} u - u_1$, we conclude on the basis of (4.3′) and (4.5) that $\nu_0(u_2) = i\sqrt{\zeta}\mu_0(u)$, whence follows (4.6′).

4.5. THEOREM. 1) *If* $\operatorname{Re}\theta < 0$, *then the operator T has precisely one eigenvalue, namely $\lambda = -\theta^2$. This eigenvalue is simple, with eigenfunction E_λ (see (4.3)). If* $\operatorname{Re}\theta \geq 0$ *or $\theta = \infty$, then T has no eigenvalues.*

2) *The resolvent set $\rho(T)$ of the operator T coincides with* $\mathbf{C}\backslash([0,\infty[\cup\{-\theta^2\})$ *if* $\operatorname{Re}\theta < 0$, *and with* $\mathbf{C}\backslash[0,\infty[$ *if* $\operatorname{Re}\theta \geq 0$. *For $\zeta \in \rho(T)$ and $f \in H$,*

$$(T - \zeta)^{-1}f(x) = \int_{\mathbf{R}^3} T(x,y;\zeta)f(y)\,dy, \tag{4.7}$$

where

$$T(x,y,\zeta) \stackrel{\text{def}}{=} \frac{\exp(i\sqrt{\zeta}|x-y|)}{4\pi|x-y|} + \frac{1}{\gamma(\zeta)}\frac{\exp(i\sqrt{\zeta}(|x|+|y|))}{(4\pi)^2|x|\,|y|}. \tag{4.7'}$$

3) *The continuous spectrum of T fills* $[0,\infty[$.

PROOF. 1) From Lemma 4.1 it follows that if λ is an eigenvalue of T, then $\lambda \notin [0,\infty[$, and the corresponding eigenfunction is E_λ. The condition $E_\lambda \in D(T)$ leads to the equality $\gamma(\lambda) = 0$ (see (4.2)) or, by (4.5′), to the equality $\sqrt{\lambda} = -i\theta$. Since $\operatorname{Im}\sqrt{\lambda} > 0$, the last equality is possible only if $\operatorname{Re}\theta < 0$.

Let $R^\theta \stackrel{\text{def}}{=} \mathbf{C}\backslash([\theta,\infty[\cup\{-\theta^2\})$ if $\operatorname{Re}\theta < 0$, and $R^\theta \stackrel{\text{def}}{=} \mathbf{C}\backslash[0,\infty[$ if $\operatorname{Re}\theta \geq 0$. Let $\zeta \in R^\theta$, $u \in D(T)$, and $(T - \zeta)u = f$. Since $T \subset S_1$, (4.6) holds by Lemma 4.4. But since $\gamma_0(u) = 0$, then by (4.6′)

$$(\hat{T} - \zeta)^{-1}\hat{f}(\xi) = \frac{1}{\xi^2 - \zeta}\left(f(\xi) - \frac{1}{(2\pi)^{3/2}}\frac{(f|E_{\bar\zeta})}{\gamma(\zeta)}\right); \tag{4.7''}$$

here, $\hat{T}$ denotes the Fourier image of T: $D(\hat{T}) \stackrel{\text{def}}{=} \Phi D(T)$ and $\hat{T}\hat{f} \stackrel{\text{def}}{=} \widehat{Tf}$, for $f \in D(T)$. On carrying out the inverse Fourier transform, from (4.7″) we get (4.7).

As before, let $\zeta \in R^\theta$, and denote by $\hat{T}_\zeta \hat{f}(\xi)$ the right-hand side of (4.7″). It is easy to see that the operator $\hat{T}_\zeta$ so defined is continuous on $H \to H$, where $R(\hat{T}_\zeta) \subset D(\hat{T})$ and $(\hat{T} - \zeta)\hat{T}_\zeta = 1_H$. Thus $\hat{T}_\zeta = (\hat{T} - \zeta)^{-1}$ and $R^\theta \subset \rho(T)$. The reverse inclusion is obtained from the following.

Let $\operatorname{Re}\zeta = \sigma > 0$ and $\operatorname{Im}\zeta = \tau \neq 0$, and let $\varepsilon > 0$ be a small number. Further, let $\hat{f}(\xi) = 1$ for $|\xi^2 - \sigma| < \varepsilon|\tau|$ and $\xi_3 > 0$, let $\hat{f}(\xi) = 0$ for $|\xi^2 - \sigma| > \varepsilon|\tau|$ and $\xi_3 > 0$, and let $\hat{f}(-\xi) = -\hat{f}(\xi)$ for all $\xi \in \mathbf{R}^3$. Since the function E_ζ is even, $(f|E_{\bar{\zeta}}) = 0$, and therefore (see (4.7″))

$$(\hat{T} - \zeta)^{-1}\hat{f}(\xi) = (\xi^2 - \zeta)^{-1} f(\xi).$$

Consequently,

$$\begin{aligned}\|(T - \zeta)^{-1} f\|^2 = \int_{|\xi^2 - \sigma| < \varepsilon|\tau|} |\xi^2 - \zeta|^{-2}\, d\xi &\geq \frac{1}{(\varepsilon^2 + 1)\tau^2} \cdot \int_{|\xi^2 - \sigma| < \varepsilon|\tau|} d\xi \\ &= \frac{1}{(\varepsilon^2 + 1)\tau^2} \|f\|^2,\end{aligned}$$

whence

$$\|(T - \zeta)^{-1}\| \geq \frac{1}{|\operatorname{Im}\tau|}. \tag{4.8}$$

Therefore, the semiaxis $[0, \infty[$ belongs to the spectrum of the operator $T = S^\theta$. From Green's formula (3.1) it follows that $T^* = S^{\bar{\theta}}$. But it has already been proved that $S^{\bar{\theta}}$ has no positive eigenvalues. Therefore, $[0, \infty[$ belongs to the continuous spectrum of T.

Note that, aside from the notation, (4.7″) coincides with the corresponding formula in [1].

In the second part of this paper we shall describe the T-Fourier transform, which diagonalizes the operator T.

BIBLIOGRAPHY

1. F. A. Berezin and L. D. Faddeev, *A remark on Schrödinger's equation with a singular potential*, Dokl. Akad. Nauk SSSR **137** (1961), 1011–1014; English transl. in Soviet Math. Dokl. **2** (1961).

2. Ya. B. Zel′dovich, *Scattering by a singular potential in perturbation theory and in the momentum representation*, Zh. Èksper. Teoret. Phys. **38** (1960), 819–824; English transl. in Soviet Phys. JETP **11** (1960).

3. F. Zirilli, *Spectral properties of some one dimensional "singular" Schrödinger Hamiltonian*, Boll. Un. Mat. Ital. B(5) **13** (1976), 355–368.

4. Yu. M. Shirokov, *Strongly singular potentials in three-dimensional quantum mechanics*, Teoret. i Mat. Fiz. **42** (1980), 45–49; English transl. in Theoret. Math. Phys. **42** (1980).

5. J. Zorbas, *Perturbation of self-adjoint operators by Dirac distributions*, J. Mathematical Phys. **21** (1980), 840–847.

6. S. Albeverio, J. E. Fenstad, and R. Høegh-Krohn, *Singular perturbations and nonstandard analysis*, Trans. Amer. Math. Soc. **252** (1979), 275–295.

7. R. A. Minlos and L. D. Faddeev, *Comment on the problem of three particles with point interactions*, Zh. Èksper. Teoret. Fiz. **41** (1961), 1850–1851; English transl. in Soviet Phys. JETP **14** (1962).

8. ——, *On the point interaction for a three-particle system in quantum mechanics*, Dokl. Akad. Nauk SSSR **141** (1961), 1335–1338; English transl. in Soviet Phys. Dokl. **6** (1961/62).

9. Yu. N. Demkov and V. N. Ostrovskiĭ, *The zero radius potential method in atomic physics*, Izdat. Leningrad. Gos. Univ. Leningrad, 1975. (Russian)

10. A. I. Baz', Ya. B. Zel′dovich, and A. M. Perelomov, *Scattering, reactions, and decay in nonrelativistic quantum mechanics*, 2nd rev. aug. ed., "Nauka", Moscow, 1971; English transl. of 1st ed., Israel Program Sci. Transls., Jerusalem, and U.S. Dept. of Commerce Clearinghouse for Federal Sci. Tech. Information, Springfield, Va., 1969.

11. V. È. Lyantse, *On some relationships among closed operators*, Dokl. Akad. Nauk SSSR **204** (1972), 542–545; English transl. in Soviet Math. Dokl. **13** (1972).

12. V. S. Vladimirov, *Equations of mathematical physics*, 3rd ed., "Nauka", Moscow, 1976; English transl. of 1st ed., Marcel Dekker, New York, 1971.

13. V. È. Lyantse, *Completely regular perturbations of a continuous spectrum*, Mat. Sb. **82 (124)** (1970), 126–156; English transl. in Math. USSR Sb. **11** (1970).

Translated by C. KELLER

Amer. Math. Soc. Transl.
(2) Vol. **144**, 1989

On the Theory of the One-Point Boundary Value Problem for the Laplace Operator. II

UDC 513.88

V. È. LYANTSE AND KH. B. MAĬORGA

The numbering of sections and formulas continues that of Part I. The bibliographical references given in Part I apply to this part as well.

5. In this section we give a description of the Fourier T-transform, i.e., the transform which diagonalizes the operator T. Henceforth we assume that([1])

$$i\theta \notin \mathbf{R}, \tag{5.1}$$

which guarantees that, for all $\xi \in \mathbf{R}^3$,

$$i\theta \pm |\xi| \neq 0. \tag{5.1$'$}$$

5.1. DEFINITION. Put

$$\Phi^\theta_\pm(x,\xi) \overset{\text{def}}{=} \frac{1}{(2\pi)^{3/2}}\left[e^{ix\xi} + \frac{i}{i\overline{\theta} \mp |\xi|}\frac{\exp(\mp i|x|\,|\xi|)}{|x|}\right]; \tag{5.2}$$

$$\Phi^\theta_\mp f(\xi) \overset{\text{def}}{=} \int_{\mathbf{R}^3} f(x)\overline{\Phi^\theta_\pm(x,\xi)}\,dx; \tag{5.3}$$

$${}^*\Phi^\theta_\pm \hat{f}(x) \overset{\text{def}}{=} \int_{\mathbf{R}^3} \hat{f}(\xi)\Phi^\theta_\pm(x,\xi)\,d\xi. \tag{5.4}$$

Later, it will be shown that the Hilbert space $H = L_2(\mathbf{R}^3)$ can be taken as the domain of $\Phi^\theta_\pm$ and ${}^*\Phi^\theta_\pm$. Meanwhile, these operators are assumed given on $L_1(\mathbf{R}^3) \cap L_2(\mathbf{R}^3)$.

5.2. THEOREM. 1) *Each of the operators* Φ^θ_+, Φ^θ_-, ${}^*\Phi^\theta_+$, *and* ${}^*\Phi^\theta_-$ *is continuous with respect to the norm of the space H, and consequently extends by*

1980 *Mathematics Subject Classification* (1985 *Revision*). Primary 35J05, 35J25; Secondary 47F05.

Translation of Teor. Funktsiĭ, Funktsional. Anal. i Prilozhen. Vyp. 39 (1983), 87–94; MR **86b**:35040. Part I is the immediately preceding paper in this volume.

([1])If $i\theta \in \mathbf{R}$, then θ^2 is a spectral singularity of the operator $T = S^\theta$.

continuity to all of H. In the previous notation for the extended operators,

$$(\Phi_{\pm}^{\theta}) = {}^*\Phi_{\pm}^{\theta}, \tag{5.5}$$

i.e., the operators $\Phi_{\pm}^{\theta}$ and ${}^*\Phi_{\pm}^{\theta}$ are mutually adjoint relative to $(\cdot|\cdot)_H$.

2) *For all $f \in D(T)$ and almost all $\xi \in \mathbf{R}^3$,*

$$\Phi_{\pm}^{\theta} T f(\xi) = \xi^2 \Phi_{\pm}^{\theta} f(\xi). \tag{5.6}$$

Conversely, if $f \in H$ and

$$\int_{\mathbf{R}^3} |\xi^2 \Phi_{\pm}^{\theta} f(\xi)|^2 \, d\xi < \infty, \tag{5.7}$$

then $f \in D(T)$.

3) *If $\operatorname{Re}\theta \geq 0$, then $\Phi_{\pm}^{\theta}$ and ${}^*\Phi_{\pm}^{\overline{\theta}}$ are mutually inverse:*

$$(\Phi_{\pm}^{\theta})^{-1} = {}^*\Phi_{\pm}^{\overline{\theta}}. \tag{5.8}$$

In particular, if $\theta \in \mathbf{R}$ and $\theta \geq 0$, then $\Phi_{\pm}^{\theta}$ is a unitary operator on $H \to H$.

4) *If $\operatorname{Re}\theta < 0$, then*

$$f = {}^*\Phi_{\pm}^{\overline{\theta}} \Phi_{\pm}^{\theta} f - 8\pi \operatorname{Re}\theta (f|E_{-\overline{\theta}^2}) E_{-\theta^2}. \tag{5.9}$$

In particular, if $\theta \in \mathbf{R}$ and $\theta < 0$, then the mapping

$$f \mapsto (\Phi_{\pm}^{\theta} f, \sqrt{-8\pi\theta}(f|E_{-\theta^2}))$$

is a unitary transformation on $H \to H \times \mathbf{C}$.

The proof of Theorem 5.2 will be carried out in the following sections.

5.3. Corollary. *For all $f, g \in H$, Parseval's identity holds:*

$$(f|g) = (\Phi_{\pm}^{\theta} f | \Phi_{\pm}^{\overline{\theta}} g) \tag{5.10}$$

if $\operatorname{Re}\theta \geq 0$, and

$$(f|g) = (\Phi_{\pm}^{\theta} f | \Phi_{\pm}^{\overline{\theta}} g) - 8\pi \operatorname{Re}\theta (f|E_{-\overline{\theta}^2})(g|E_{-\theta^2}) \tag{5.10$'$}$$

if $\operatorname{Re}\theta < 0$.

6. In this auxiliary section, we discuss some properties of the operator Q_τ defined by

$$Q_\tau \hat{f}(\xi) = \frac{1}{2\pi i} \int_{\mathbf{R}^3} \frac{\hat{f}(\eta)\, d\eta}{\eta^2 - \xi^2 - i\tau}, \qquad \xi \in \mathbf{R}^3, \ \tau \in \mathbf{R}\setminus\{0\}. \tag{6.1}$$

6.1. Definition. Let ω_ρ be Lebesgue measure on the sphere $\{\xi \in \mathbf{R}^3 : |\xi| = \rho\}$, and let

$$\Omega \hat{f}(\rho) \stackrel{\text{def}}{=} \int_{|\xi|=\rho} \hat{f}(\xi) \omega_\rho(d\xi); \tag{6.2}$$

$$R\hat{f}(\rho) \stackrel{\text{def}}{=} \frac{1}{2\sqrt{\rho}} \Omega \hat{f}(\sqrt{\rho}), \qquad \rho > 0. \tag{6.3}$$

In addition, for all $\alpha \in \mathbf{R}$, denote by X^α the Hilbert space of functions $\hat{f}\colon \mathbf{R}^3 \to \mathbf{C}$ corresponding to the norm

$$\|\hat{f}\|_{X^\alpha}^2 \stackrel{\text{def}}{=} \int_{\mathbf{R}^3} \xi^{2\alpha}|\hat{f}(\xi)|^2\,d\xi. \tag{6.4}$$

6.2. LEMMA.[2]

$$R \in B(X^{1/2}; L_2(0,\infty)). \tag{6.5}$$

PROOF. Since

$$|\Omega\hat{f}(\rho)|^2 \le 4\pi\rho^2 \int_{|\xi|=\rho} |\hat{f}(\eta)|^2\omega_\rho(d\eta),$$

we have

$$\int_0^\infty |R\hat{f}(\rho)|^2\,d\rho \le 2\pi \int_{\mathbf{R}^3} |\xi|\,|\hat{f}(\xi)|^2\,d\xi. \tag{6.5$'$}$$

6.3. DEFINITION. For all $\varphi \in L_2(\mathbf{R})$, $\sigma \in \mathbf{R}$, and $\tau \in \mathbf{R}\backslash\{0\}$, put

$$P_\tau\varphi(\sigma) \stackrel{\text{def}}{=} \frac{1}{2\pi i}\int_{\mathbf{R}} \frac{\varphi(s)\,ds}{s-(\sigma+i\tau)}. \tag{6.6}$$

The following assertion is well known (see [13]).

6.4. LEMMA. *For all $\tau \in \mathbf{R}\backslash\{0\}$, $P_\tau \in B(L_2(\mathbf{R}))$, and the strong limits $P_\pm \stackrel{\text{def}}{=} \operatorname{s-lim}_{\tau\downarrow 0} P_{\pm\tau}$ exist. The operators P_+ and $-P_-$ are mutually complementary orthogonal projections of the space $L_2(\mathbf{R})$; in particular, $P_+ - P_- = 1_{L_2(\mathbf{R})}$.*

6.5. DEFINITION. For all $\psi \in L_2(0,\infty)$ put $W\psi(\xi) \stackrel{\text{def}}{=} \psi(\xi^2)$, $\xi \in \mathbf{R}^3$. It is easy to see that (see (6.4))

$$\|W\psi\|_{X^{-1/2}} = \sqrt{2\pi}\|\psi\|_{\Delta_2(0,\infty)}. \tag{6.7}$$

6.6. ASSERTION. *For the operator Q_τ (see (6.1)) the inclusion $Q_\tau \in B(X^{1/2}; X^{-1/2})$, $\tau \neq 0$, holds. The strong limits $Q_\pm \stackrel{\text{def}}{=} \operatorname{s-lim}_{\tau\downarrow 0} Q_{\pm\tau} \in B(X^{1/2}; X^{-1/2})$ exist.*

PROOF. Let l be the operator $L_2(0,\infty) \to L_2(\mathbf{R})$ of extension by zero to $]-\infty, 0[$, and l^* be its adjoint operator (restriction from $\mathbf{R}$ to $]0,\infty[$). It is easily checked that

$$Q_\tau = Wl^*P_\tau lR. \tag{6.8}$$

Since $\|l\| = \|l^*\| = 1$, our assertion follows from Lemmas 6.2 and 6.4 and relation (6.7).

6.7. REMARK. Formally, $Q_\pm$ is an integral operator:

$$Q_\pm\hat{f}(\xi) = \int_{\mathbf{R}^3} Q_\pm(\xi,\eta)\hat{f}(\eta)\,d\eta.$$

---

(2) For arbitrary normed spaces X and Y, we denote by $B(X;Y)$ the space of linear, everywhere defined, continuous operators $X \to Y$, with the usual operator norm.

Its kernel is the "generalized function"

$$Q_\pm(\xi,\eta) = \frac{1}{2\pi i}\cdot\frac{1}{\eta^2-\xi^2\mp i0}.$$

7. To prove Theorem 5.2, we shall use the following notation: for all $f\in H$, $\xi\in\mathbf{R}^3$, and $\tau\in\mathbf{R}\backslash\{0\}$,

$$L_\tau f(\xi) = \hat{L}_\tau\hat{f}(\xi) \overset{\text{def}}{=} -\frac{1}{(2\pi)^{3/2}}\frac{1}{\gamma(\xi^2+i\tau)}(f|E_{\xi^2-i\tau}); \tag{7.1}$$

$$\Phi_\tau f(\xi) \overset{\text{def}}{=} (1+\hat{L}_\tau)\hat{f}(\xi). \tag{7.2}$$

7.1. LEMMA. *If $f\in D(T)$, then*

$$\Phi_\tau Tf(\xi) = \xi^2\Phi_\tau f(\xi) + i\tau\hat{L}_\tau\hat{f}(\xi). \tag{7.3}$$

PROOF. If $f\in D(T)$, then, by (4.6′),

$$((T-\zeta)f|E_{\bar\zeta}) = -4\pi\gamma(\zeta)\mu_0(f),$$

or $\gamma_0(f)=0$. Consequently (see (7.1)),

$$\hat{L}_\tau\hat{T}\hat{f}(\xi) = (\xi^2+i\tau)\hat{L}_\tau\hat{f}(\xi) + \sqrt{2/\pi}\mu_0(f).$$

Since $\hat{T}\hat{f}(\xi) = \xi^2\hat{f}(\xi) - \sqrt{2/\pi}\mu_0(f)$, we get (7.3).

7.2. LEMMA. *If $f\in H\cap L_1(\mathbf{R}^3)$, then for all $\xi\in\mathbf{R}^3$ the limits*

$$\Phi_\pm f(\xi) \overset{\text{def}}{=} \lim_{\nu\downarrow 0}\Phi_{\mp\tau}f(\xi) \tag{7.4}$$

exist, and (*see* (5.3))

$$\Phi_\pm f(\xi) = \Phi_\pm^\theta f(\xi). \tag{7.4′}$$

PROOF. From (4.3), (4.5′), and (7.1) it follows that

$$\hat{L}_\tau\hat{f}(\xi) = \frac{1}{(2\pi)^{3/2}}\frac{i}{i\theta+\sqrt{\xi^2+i\tau}}\int_{\mathbf{R}^3} f(x)\frac{\exp(i\sqrt{\xi^2+i\tau}|x|)}{|x|}\,dx. \tag{7.1′}$$

Our assertion follows from (7.1′) and (7.2).

7.3. COROLLARY. *If $f\in D(T)\cap L_1(\mathbf{R}^3)$ and $Tf\in L_1(\mathbf{R}^3)$, then for all $\xi\in\mathbf{R}^3$*

$$\Phi_\pm^\theta Tf(\xi) = \xi^2\Phi_\pm^\theta f(\xi). \tag{7.5}$$

For the proof it suffices to pass to the limit in (7.3) as $\tau\to\pm 0$.

Note that Corollary 7.3 is close to statement 2) of Theorem 5.2, although until now we have had insufficient information on the range of $\Phi_\pm^\theta$. In order to fill in this gap, we observe that

$$\hat{L}\hat{f}(\xi) = -\frac{1}{\pi}\frac{1}{i\theta+\sqrt{\xi^2+i\tau}}Q_\tau\hat{f}(\xi), \tag{7.1″}$$

where Q_τ is defined by (6.1).

7.4. Lemma. *For all sufficiently small* $\tau \in \mathbf{R}\backslash\{0\}$, $\hat{L}_\tau \in B(X^{1/2})$. *The strong limits*

$$\hat{L}_\pm \stackrel{\text{def}}{=} \operatorname*{s-lim}_{\tau\downarrow 0} \hat{L}_{\pm\tau} \in B(X^{1/2})$$

exist.

Proof. By condition (5.1), $i\theta + \sqrt{\xi^2 + i\tau} \neq 0$ for $\xi \in \mathbf{R}^3$. Therefore, the operator of multiplication by the function

$$\xi \mapsto (i\theta + \sqrt{\xi^2 + i\tau})^{-1}$$

belongs to $B(X^{-1/2}; X^{1/2})$. Consequently, our assertion follows from 6.6.

7.5. Corollary. *For a suitable choice of domain of the operators* $\Phi^\theta_\pm$,

$$\Phi_\pm = \operatorname*{s-lim}_{\tau\downarrow 0} \Phi_{\pm\tau} \in B(H^{1/2}; X^{1/2}). \tag{7.6}$$

This follows from (7.2) and Lemma 7.4, since $\|\hat{f}\|_{X^{1/2}} \leq \|f\|_{H^{1/2}}$.

7.6. Remark. Let $\hat{\Phi}_\pm$ denote the Fourier image of the operator $\Phi_\pm$. Taking (7.1), (7.2), and (4.3′) into account, we conclude that $\hat{\Phi}_\pm$ can be treated formally as the integral operator

$$\hat{\Phi}_\pm \hat{f}(\xi) = \int_{\mathbf{R}^3} \hat{f}(\eta)\overline{\hat{\Phi}_\pm(\xi,\eta)}\, d\eta$$

with "kernel"

$$\hat{\Phi}_\pm(\xi,\eta) = \delta(\xi, -\eta) \times \frac{1}{2\pi^2(i\overline{\theta} \mp |\xi|)} \times \frac{1}{\eta^2 - \xi^2 \pm i\theta}.$$

7.7. Remark. From (4.4), (4.5′), (7.1), (7.2), (7.4), and (7.4′) it follows that

$$\Phi^\theta_+ E_\zeta(\xi) = -\frac{1}{(2\pi)^{3/2}}\left[\frac{1}{\xi^2 - \zeta} - \frac{1}{i\theta + |\xi|}\frac{1}{\sqrt{\xi} + |\xi|}\right]. \tag{7.7}$$

In particular, if $\operatorname{Re}\theta < 0$, then

$$\Phi^\theta_+ E_{-\theta^2} = 0. \tag{7.8}$$

7.8. Remark. From (5.3) and (5.4) it is easy to see that the operators $\Phi^\theta_\pm$ and ${}^*\Phi^\theta_\pm$ are formally mutually adjoint relative to $(\cdot|\cdot)_H$. Since the dual to X^s (to H^s) with respect to $(\cdot|\cdot)_H$ is the space X^{-s} (the space H^{-s}), then, taking (7.6) into account, it can be assumed that, for a suitable choice of domain,

$${}^*\Phi^\theta_\pm \in B(X^{-1/2}; H^{-1/2}). \tag{7.9}$$

An analogous observation relates to the operator $\hat{L}_\pm$ defined in Lemma 7.4. Later it will be convenient to take note of the dependence of this operator on θ and denote it by $\hat{L}^\theta_\pm$. If ${}^*\hat{L}^\theta_\pm$ is the operator adjoint to $\hat{L}^\theta_\pm$ relative to $(\cdot|\cdot)_H$, then for suitable choice of domain we have, by Lemma 7.4,

$${}^*\hat{L}^\theta_\pm \in B(X^{-1/2}). \tag{7.10}$$

8. To complete the proof of Theorem 5.2, we shall need the following observation, which is obvious by Lemma 6.4.

8.1. REMARK. Let $\varphi \in L_2(0,\infty)$, $\zeta \in \mathbf{C}\backslash[0,\infty[$, and

$$M\varphi(\zeta) \stackrel{\text{def}}{=} \frac{1}{2\pi i}\int_0^\infty \frac{\varphi(s)\,ds}{s-\zeta}. \tag{8.1}$$

Then $M\varphi$ is holomorphic on $\mathbf{C}\backslash[0,\infty[$, and

$$|M\varphi(\zeta)| \le \frac{1}{2\sqrt{\pi}}\|\varphi\|_{L_2(0,\infty)}\frac{1}{|\operatorname{Im}\zeta|^{1/2}}. \tag{8.2}$$

Moreover, the limits

$$M_\pm\varphi(\sigma) = \underset{\tau\downarrow 0}{\text{l.i.m.}}\, M\varphi(\sigma+i\tau), \qquad \sigma \in \mathbf{R},$$

exist in the norm of $L_2(\mathbf{R})$, and for almost all $\sigma \in \mathbf{R}$,

$$M_+\varphi(\sigma) - M_-\varphi(\sigma) = \varphi(\sigma). \tag{8.3}$$

The following assertion is fundamental.

8.2. LEMMA. *For all $f \in H^{1/2}$, if* $\operatorname{Re}\theta \ge 0$, *then*

$$\Phi f = \hat{f} = (1 + {}^*\hat{L}^\theta_\pm)\Phi^\theta_\pm f, \tag{8.4}$$

and if $\operatorname{Re}\theta < 0$, *then*

$$\Phi f = \hat{f} = (1 + {}^*\hat{L}^{\overline{\theta}}_\pm)\Phi^\theta_\pm f - 8\pi\operatorname{Re}\theta(f|E_{-\overline{\theta}^2})E_{-\theta^2}, \tag{8.5}$$

where ${}^*\hat{L}^\theta_\pm$ *is the operator described in Remark* 7.8.

PROOF. It is easy to see that

$$\Phi^\theta_\pm f(\xi) = \hat{f}(\xi) + \frac{1}{(2\pi)^3}\frac{1}{\gamma\pm(\xi^2)}P_\pm lR\hat{f}(\xi^2), \tag{8.6}$$

where (see (4.5′))

$$\gamma_\pm(\xi^2) = \lim_{\tau\downarrow 0}\gamma(\xi^2 \pm i\tau) = \frac{(i\theta \pm |\xi|)}{4\pi i}, \tag{8.7}$$

and the definitions of the operators $P_\pm$, l, and R are formulated in §6. Let us apply the operator R to both sides of (8.6). Putting, for brevity,

$$g_\pm = R\Phi^\theta_\pm f, \qquad h = R\hat{f} = R\Phi f, \tag{8.8}$$

we find that

$$g_\pm(\rho) = h(\rho) + \frac{1}{2\pi}\frac{i\sqrt{\rho}}{\gamma\pm(\rho)}P_\pm lh(\rho), \qquad \rho > 0. \tag{8.9}$$

Note that

$$\gamma_+(\rho)g_+(\rho) = \gamma_-(\rho)g_-(\rho), \qquad \rho > 0. \tag{8.10}$$

Indeed, by (8.7),

$$\gamma_+(\rho) - \gamma_-(\rho) = \frac{\sqrt{\rho}}{2\pi i}, \qquad \rho > 0, \tag{8.11}$$

and therefore (8.10) follows from (8.9) and the equality $P_+ - P_- = 1$. In connection with (8.10), we shall henceforth consider only the + index.

Let $\zeta \in \rho(T)$ (see 4.5). Put

$$F(\zeta) \overset{\text{def}}{=} \frac{1}{2\pi i}\frac{1}{\gamma s}\int_0^\infty \frac{h(s)\,ds}{s-\zeta} = \frac{1}{\gamma(s)} P_\tau l h(\rho), \tag{8.12}$$

where $\sigma, \tau \in \mathbf{R}$ and $\sigma + i\tau = \zeta$. Denote, also,

$$G(\zeta) \overset{\text{def}}{=} \frac{1}{2\pi i}\int_0^\infty \frac{g_+(s)\,ds}{\gamma_-(s)(s-\zeta)}. \tag{8.13}$$

By (5.1), $\gamma_\pm(s) \neq 0$ for $s \in \mathbf{R}$, and thus Remark 8.1 can be applied to the functions F and G. Consequently, $\gamma_+(\sigma)F_+(\sigma) - \gamma_-(\sigma)F_-(\sigma) = h(\sigma)$, $\sigma > 0$, and $G_+(\sigma) - G_-(\sigma) = g_+(\sigma)/\gamma_-(\sigma)$, $\sigma > 0$, where the indices + and − denote the upper and lower boundary values on the cut $[0, \infty[$. Now, from (8.9), taking (8.11) into account, we get

$$F_+(\sigma) - G_+(\sigma) = F_-(\sigma) - G_-(\sigma). \tag{8.14}$$

Hence $Y \overset{\text{def}}{=} F - G$ extends holomorphically to $[0, \infty[$. Indeed, from (8.12) and (8.13) it is clear that Y is holomorphic on $\rho(T)$, and therefore

$$\forall \zeta \in \rho(T) \qquad Y(\zeta) = \frac{1}{2\pi i}\oint \frac{Y(z)\,dz}{z-\zeta}; \tag{8.15}$$

if $\operatorname{Re}\theta \geq 0$, the integration is carried out along the contour indicated in the figure, while in the case $\operatorname{Re}\theta < 0$ and $|\theta|^2 \leq |\zeta|$ a small circle should be added to the pictured contour, with center $-\theta^2$, oriented counterclockwise (since $-\theta^2$ is a pole of F). Since $Y_+(\sigma) = Y_-(\sigma)$ for almost all $\sigma > 0$ (by (14)), taking into account that $|Y(\zeta)| \leq \text{const}\,|\operatorname{Im}\zeta|^{-1/2}$ near $[0, \infty[$ (see (8.2)), we see that the contour indicated in the figure can be replaced by a circle with center at zero and radius greater than $|\zeta|$. This will give the required holomorphic extension.

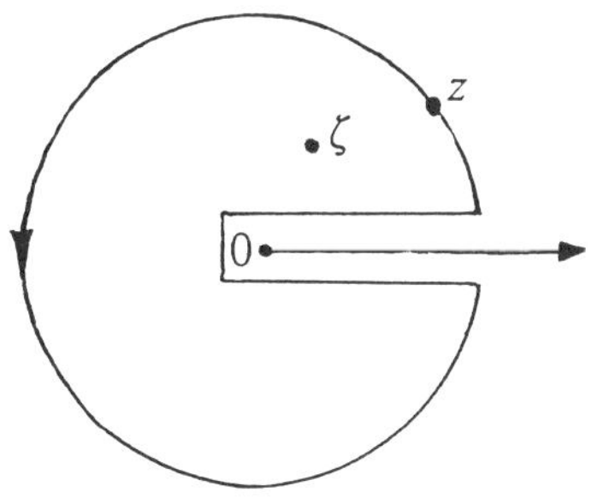

By (8.2) we have an estimate of the form

$$\left| Y(\zeta) - \frac{c'}{\zeta + \theta^2} \right| \leq \frac{\text{const}}{|\operatorname{Im}\zeta|^{1/2}},$$

where c' is the residue of the function F at the point $\zeta = -\theta^2$; if $\operatorname{Re}\theta \geq 0$ then $c' = 0$. Applying the Cauchy inequality for the power series coefficients, we get $Y(\zeta) = c'(\zeta + \theta^2)^{-1}$, i.e., $F(\zeta) = G(\sigma) + c'/(\zeta + \theta^2)$. Consequently, taking (8.12) into account, we can rewrite (8.6) as

$$\Phi^\theta_\pm f(\xi) = \hat{f}(\xi) + \frac{1}{(2\pi)^3}\left[G_+(\xi^2) + \frac{c'}{\zeta + \theta^2}\right]. \tag{8.16}$$

Applying in succession (8.13), (8.8), and (6.3), we find that

$$G_+(\xi^2) = \frac{1}{2\pi i}\lim_{\tau\downarrow 0}\int_{\mathbf{R}^3}\frac{\Phi^\theta_+ f(\eta)\,d\eta}{\gamma_-(\eta^2)(\eta^2 - \xi^2 - i\tau)} = (2\pi)^3 i^* \hat{L}^{\overline{\theta}}_+ \Phi^\theta_+ f(\xi).$$

Hence, from (8.16) it follows that

$$\hat{f}(\xi) = \Phi^\theta_- f(\xi) + {}^*\hat{L}^{\overline{\theta}}_+ \Phi^\theta_+ f(\xi) + c\hat{E}_{-\theta^2}(\xi), \tag{8.17}$$

where $c = 0$ for $\operatorname{Re}\theta \geq 0$. With this, (8.4) is proved. Let us do the calculation in the case $\operatorname{Re}\theta < 0$. By (7.8),

$$((1 + {}^*\hat{L}^{\overline{\theta}}_+)\Phi^\theta_+ f|\hat{E}_{-\theta^2}) = (\Phi^\theta_+ f|\Phi^{\overline{\theta}}_+ E_{-\theta^2}) = 0.$$

Since, by (4.4),

$$(E_{-\theta^2}|E_{-\overline{\theta}^2}) = -1/(8\pi \operatorname{Re}\theta),$$

from (8.17) we obtain $(f|E_{-\theta^2}) = -c/(8\pi \operatorname{Re}\theta)$, whence follows (8.5).

8.3. COROLLARY. *For all $f, g \in H^{1/2}$, Parseval's identity* (5.10) *holds for* $\operatorname{Re}\theta \geq 0$, *and* (5.10′) *holds for* $\operatorname{Re}\theta < 0$.

Indeed, from (8.4) it follows that

$$\begin{aligned}(f|g) = (\Phi f|\Phi g) &= ((1 + {}^*\hat{L}^\theta_\pm)\Phi^\theta_\pm f|\Phi g)\\ &= (\Phi^\theta_\pm f|(1 + \hat{L}^{\overline{\theta}}_\pm)\Phi g) = (\Phi^\theta_\pm f|\Phi^\theta_\pm g);\end{aligned}$$

the argument is analogous when $\operatorname{Re}\theta < 0$.

8.4. COROLLARY. *The operator $\Phi^\theta_\pm$ is continuous relative to $\|\cdot\|_H$.*

PROOF. Assume, first, that $\theta \in \mathbf{R}$. Then for $\theta \geq 0$ it is evident from (5.10) that the operator $\Phi^\theta_\pm$ is isometric on $H \to H$, while for $\theta < 0$ the operator $\Phi^\theta_\pm$ differs from an isometry by a $\|\cdot\|_H$-continuous one-dimensional operator. Hence we conclude that the operator $\hat{L}^\theta_\pm$ is $\|\cdot\|_H$-continuous, since $\Phi^\theta_\pm = (1 + \hat{L}^\theta_\pm)\Phi$. However, for arbitrary $\theta_1, \theta_2 \in \mathbf{C}$ the operators $\hat{L}^{\theta_1}_\pm$ and $\hat{L}^{\theta_2}_\pm$ differ by the operator of multiplication by the function

$$\xi \mapsto (i\theta_1 + |\xi|)(i\theta_2 + |\xi|)^{-1},$$

which is bounded (for $i\theta_j \notin \mathbf{R}$). Therefore the operator $\hat{L}^\theta_\pm$, and along with it $\Phi^\theta_\pm$, is $\|\cdot\|_H$-continuous for complex θ as well (such that $i\theta \in \mathbf{R}$).

8.5. COROLLARY. *For all $f, g \in H$, Parseval's identity* (5.10) *holds for* $\operatorname{Re}\theta \geq 0$ *and* (5.10′) *for* $\operatorname{Re}\theta < 0$, *since $H^{1/2}$ is dense in H, and $\Phi^\theta_\pm$ can be extended by continuity to all of H.*

With this, the proof of Theorem 5.2 is practically complete. It may be further noted that, since the operators $\Phi^\theta_\pm$ and ${}^*\Phi^\theta_\pm$ are mutually adjoint with respect to $(\cdot|\cdot)_H$ on dense subspaces of H, and for a suitable choice of domain of $\Phi_{\pm}^{\theta} \in B(H)$, then in this sense $^*\Phi^{\theta} \in B(H)$.

Translated by C. KELLER

Amer. Math. Soc. Transl.
(2) Vol. **144**, 1989

On the Cohomology of von Neumann Algebras

UDC 517.39+517.91

G. N. ZHOLTKEVICH*

§1. Introduction

This paper is devoted to the proof of the equivalence of the properties of amenability and injectivity for arbitrary von Neumann algebras. This problem has been solved completely by Connes [1] for von Neumann algebras with a separable predual. We are going to show that for general finite and semifinite von Neumann algebras, the problem can be reduced to the case considered in [1]. Takesaki's duality theory [2] enables us to solve the problem for algebras of type III as well. We shall prove that the property of injectivity for von Neumann algebras implies amenability, since the implication "amenability implies injectivity" has already been proved in [1] without restrictions on the predual of the von Neumann algebra.

§2. Basic definitions and notation

Recall that a normal dual Banach bimodule over a von Neumann algebra M is, by definition, a Banach space X which is dual to some Banach space X_*, is an M-bimodule and possesses the following properties.

1) $||x\xi y|| \leq K||x|| \, ||y|| \, ||\xi||$ $(x, y \in M,\ \xi \in X)$.

2) The operators $\xi \to x\xi$ and $\xi \to \xi x$ are $\sigma(x, x_*)$-continuous.

3) The functionals of the form $\langle \eta, x\xi \rangle$ and $\langle \eta, \xi x \rangle$, defined on M, where $\eta \in X_*$ and $\xi \in X$, are normal.

1980 *Mathematics Subject Classification* (1985 *Revision*). Primary 46L10; Secondary 46L35.
Translation of Teor. Funktsiĭ, Funktsional. Anal. i Prilozhen. Vyp. 38 (1982), 122–125; MR **84c**: 46064.

**Editor's note.* The author's name was misprinted "Zholkevich" on the original paper.

A normal differentiation on M with values in a dual normal Banach M-bimodule X is, by definition, a linear mapping $D: M \to X$ which is $\sigma(X, X_*)$-weakly continuous and satisfies the condition $D(xy) = D(x)y + xD(y)$.

A normal differentiation D is called *inner* if there is $\xi \in X$ such that $D(x) = x\xi - \xi x$.

A von Neumann algebra is said to be *amenable* if for any normal dual Banach M-bimodule X, all differentiations on M with values in X are inner.

§3. Amenable groups and von Neumann algebras

In this section we shall prove three lemmas which show the connection between the notions of amenability for groups and von Neumann algebras.

LEMMA 3.1. *Let M be a von Neumann algebra and let G be an amenable subgroup of the unitary group of M, such that $G'' = M$. Then M is amenable.*

PROOF. Let X be any dual normal Banach M-bimodule and let D be a normal differentiation in X.

It is obvious that $u^*D(u) \in L^\infty(G, X)$, the space of bounded X-valued functions on G, for $u \in G$. Applying μ of Lemma 3.3 from [3] to the equality $w(uw)^*D(uw) = u^*D(u)w + D(w)$, where $u, w \in G$, we obtain $w\xi = \xi w + D(w)$, $\xi = \mu(u^*D(u))$.

Since $G'' = M$, we obtain that $D(x) = x\xi - \xi x$ for any $x \in M$, i.e., M is amenable.

LEMMA 3.2. *Let N be an amenable von Neumann algebra and let G be an amenable group acting on N by automorphisms. Then $M = W^*(N, G)$ is amenable.*

PROOF. Let X be any dual normal M-bimodule and let D be a normal differentiation from M into X. Since N is amenable, we may assume that $D|N = 0$. We consider the function $\varphi \in l^\infty(G, X)$ given by $\varphi(g) = \lambda_g^* D(\lambda_g)$, and write $\xi = \mu(\varphi)$.

If $y \in N$, then

$$\begin{aligned} y\lambda_g^* D(\lambda_g) &= \lambda_g^*(\lambda_g y \lambda_g^* D(\lambda_g)) = \lambda_g^*[\lambda_g y \lambda_g^* \cdot D(\lambda_g) + D(\lambda_g y \lambda_g^*)\lambda_g] \\ &= \lambda_g^* D(\lambda_g y) = \lambda_g^* D(\lambda_g) y \cdot y\varphi(g) = \varphi(g) y. \end{aligned} \tag{1}$$

Let $h \in G$. Then

$$\begin{aligned} \lambda_h^* \varphi(g) &= \lambda_h^* \lambda_g^* D(\lambda_g) = \lambda_{gh}^* D(\lambda_{gh} \lambda_h^*) \\ &= \lambda_{gh}^* D(\lambda_{gh}) \lambda_h^* + D(\lambda_h^*) \lambda_h^* \varphi(g) = \varphi(gh) \lambda_h^* + D(\lambda_h^*). \end{aligned} \tag{2}$$

Applying μ to (1) and (2), we obtain $y\xi = \xi y$ and $D(\lambda_h^*) = \lambda_h^* \xi - \xi \lambda_h^*$, and so $D(x) = x\xi - \xi x$, since $M = W^*(N, (\lambda_g))$, i.e., M is amenable.

LEMMA 3.3. *Let M be an injective semifinite von Neumann algebra acting in a separable Hilbert space H. Then in the unitary group of M there is an amenable subgroup G such that $G'' = M$.*

PROOF. In the center $Z(M)$ we choose orthogonal projections E_k such that $\sum E_k = 1$, $E_1 M$ is an algebra of type II_1, $E_2 M$ is an algebra of type II_∞, $E_3 M$ is an algebra of type I_∞, and $E_k M$ is an algebra of type I_{n_k} for $k \geq 4$ (see [4], pp. 98, 122, and 240). It is obvious that all the $E_k M$ are injective and act in the separable spaces $E_k H$, $H = \bigoplus E_k H$.

By [4] (pp. 239–240), $E_k M \simeq B(H_k) \oplus C_k$ for $k \geq 3$, where C_k is a commutative von Neumann algebra and H_k is a Hilbert space of dimension n_k.

Let U_k be an amenable unitary subgroup of $B(H_k)$ such that $U_k'' = B(H_k)$, and let V_k be the unitary group of C_k. Then $G_k = \{u' \otimes u'' | u' \in U_k, u'' \in V_k\}$ generates $E_k M$, and is obviously amenable.

Now we consider the central decomposition $E_1 M \simeq \int \bigoplus M_t d\mu(t)$. According to [5], Theorem A, all factors M_t are injective, and consequently ([6], 7.1, 7.2, 7.5, 7.7, 7.8), they are hyperfinite.

Therefore, $E_1 M \simeq R_1 \otimes C_1$, where C_1 is a commutative von Neumann algebra and R_1 is a hyperfinite factor of type II_1 ([4], p. 179).

Since R_1 is generated by an amenable unitary subgroup, $E_1 M$ is generated by some amenable unitary subgroup of C_1. Similarly, $E_2 M = C_2''$, where C_2 is amenable.

Let the group W_n act in $E_k H$ as G_k for $k \leq n$ and as the identity for $k > n$. Then $W_n \subset W_{n+1}$, and by [7], Theorem I.2.7, $G = \bigcup W_n$ is amenable and $G'' = M$.

§4. Amenability of injective von Neumann algebras

In this section we show that every injective von Neumann algebra is amenable. For von Neumann algebras of type I, this has been proved in [3]. Therefore, we shall carry out the proof for von Neumann algebras of types II_1, II_∞, and III.

4.1. *The case of a type* II_1 *algebra.*

LEMMA 4.1.1. *Let M be an injective weakly separable von Neumann algebra of type* II_1. *The unitary group of M contains an amenable subgroup G such that* $G'' = M$.

PROOF. Let $T(M)$ be the set of finite normal traces on M. By the finiteness of M, this set is separating. We denote by $T_0(M)$ a separating subset of traces with mutually orthogonal supports.

For every $\tau \in T_0(M)$, we denote by π_τ the normal representation of M constructed from the trace τ. Since $T_0(M)$ is separating, the algebra $N = \bigoplus_{\tau \in T_0(M)} \pi_\tau(M)$ is isomorphic to M.

Each of the algebras $\pi_\tau(M)$ acts in a separable Hilbert space H_τ, and by Lemma 3.3 it is generated by an amenable subgroup G_τ of its unitary group. We consider the set Λ of finite subsets of $T_0(M)$, and for every $F \in \Lambda$ we construct the group W_F in N that acts as G_τ in H_τ for $\tau \in F$ and as the identity for $\tau \notin F$. Then $W = \bigcup W_F$ is amenable ([7], Theorem I.2.7) and obviously generates N.

The group $G \subset M$ corresponding to W satisfies all requirements of the lemma.

THEOREM 4.1.2. *Let M be an injective von Neumann algebra of type* II_1. *Then M is amenable.*

PROOF. We consider the set Λ (by inclusion) of all countable subgroups of the unitary group of M. For every $G \in \Lambda$, we define $M_G = G''$. The algebras M_G are weakly separable, and $M = \bigcup M_G$.

Since M is finite and injective, all the M_G are injective. By Lemma 4.1.1, M_G is generated by an amenable subgroup W_G of its unitary group.

Let X be any normal dual M-bimodule and let D be a normal differentiation in X. By Lemma 3.1, $(D|M_G)(x) = x\xi_G - \xi_G x$, $\xi_G \in X$, and it is obvious from the proof of Lemma 3.1 that $||\xi_G|| \leq ||D||$.

By the $\sigma(X, X_*)$-compactness of the unit ball of X, in $\{\xi_G\}$ there is a $\sigma(X, X_*)$-convergent subnet $\xi_\alpha \to \xi$. Hence $D(x) = x\xi - \xi x$, i.e., M is amenable.

4.2. *The case of algebras of type* II_∞

THEOREM 4.2.1. *Let M be an injective von Neumann algebra of type* II_∞. *Then M is amenable.*

PROOF. By [4], p. 99, there exists a faithful normal semifinite trace τ on M. Let $N_\tau = \{x \in M | \tau(x^*x) < \infty\}$.

We consider the net Λ of countable subsets of N_τ. For $F \in \Lambda$, we construct the algebra $M_F = (F, F^*, 1)''$. All the M_F are of type II_∞ and weakly separable. Therefore, the faithful representation of M_F constructed from τ acts in a separable Hilbert space. Since M is semifinite and injective and M_F is semifinite, M_F is injective. By Lemma 3.3, M_F is generated by an amenable subgroup of its unitary group. Let X be any normal dual M-bimodule and let D be a normal differentiation in X. As in Theorem 4.1.2, $(D|M_F)(x) = x\xi_F - \xi_F x$, where $||\xi_F|| \leq ||D||$, i.e., $D(x) = x\xi - \xi x$, $x \in \bigcup M_F$, and ξ is the $\sigma(X, X_*)$-limit of $\{\xi_F\}$. On the other hand, since $(\bigcup M_F)'' = M$, we have $D(x) = x\xi - \xi x$, i.e., M is amenable.

4.3. *The case of algebras of type* III

THEOREM 4.3.1. *Let M be an injective von Neumann algebra of type* III. *Then M is amenable.*

PROOF. Let $(N, (\theta_t)_{t \in R})$ be a continuous decomposition of M [2], where N is an injective von Neumann algebra of type II_∞ and $(\theta_t)_{t \in R}$ is a one-parameter group of automoprhisms of N. By Theorem 4.2.1, N is amenable. Since M is isomorphic to $W^*(N, (\theta_t)_{t \in R})$, M is amenable by Lemma 3.2.

4.4. *The general case.*

THEOREM 4.4.1. *A von Neumann algebra M is amenable if and only if it is injective.*

PROOF. That amenability implies injectivity has been proved in [1].

Let M be an injective von Neumann algebra. By [4], pp. 98 and 122, $M = M_1 \oplus M_2 \oplus M_3 \oplus M_4$, where M_1 is an algebra of type I, M_2 is an algebra of type II_1, M_3 is an algebra of type II_∞, and M_4 is an algebra of type III. All the M_i are injective, and so they are amenable, as has been shown already. Let X be a normal dual Banach M-bimodule, $E_i \in Z(M)$ the projections corresponding to the M_i, and D a normal differentiation in X.

By the amenability of M_i, we have $D(E_i x) = (E_i x)\xi_i - \xi_i(E_i x)$, $i = 1, \ldots, 4$. Since D is a differentiation, $D(1) = 0$, i.e.,

$$\sum_{i=1}^{4} E_i \xi_i = \sum_{i=1}^{4} \xi_i E_i.$$

Therefore $D(x) = x\xi - \xi x$, $\xi = \sum_1^4 E_i \xi_i$, and so M is amenable.

BIBLIOGRAPHY

1. A. Connes, *On the cohomology of operator algebras*, J. Functional Anal. **28** (1978), 248–253.

2. Masamichi Takesaki, *Duality for crossed products and the structure of von Neumann algebras of type* III, Acta Math. **131** (1973), 249–310.

3. Richard V. Kadison and John R. Ringrose, *Cohomology of operator algebras*. I: *Type* I *von Neumann algebras*, Acta Math. **126** (1971), 227–243.

4. Jacques Dixmier, *Les algèbres d'opérateurs dans l'espace hilbertien* (*algèbres de von Neumann*), 2nd ed., Gauthier-Villars, Paris, 1969.

5. Man Duen Choi and Edward G. Effros, *Separable nuclear C^*-algebras and injectivity*, Duke Math. J. **43** (1976), 309–322.

6. A. Connes, *Classification of injective factors. Cases* II_1, II_∞, III_λ, $\lambda \neq 1$, Ann. of Math. (2) **104** (1976), 73–115.

7. Frederick P. Greenleaf, *Invariant maps on topological groups and their applications*, Van Nostrand Reinhold, New York, 1969.

Translated by J. SZÜCS

Amer. Math. Soc. Transl.
(2) Vol. **144**, 1989

On Approximate Algebraic Stability Criteria

UDC 512.8.517.9

N. A. PUSTOVOĬTOV

Abstract. The author gives a detailed analysis and develops in a new way an earlier [7] matrix algorithm to construct approximate stability criteria for linear autonomous differential equations whose coefficients depend on a vector-valued parameter.

1. The first mathematical formulation of the problem of a stability criterion for a (linear autonomous) differential equation arising in connection with the study of centrifugal regulators of steam engines is due to Maxwell. In a meeting of the London Mathematical Society in 1868 he posed the problem of estimating the signs of the real parts of the roots of an algebraic equation directly in terms of its coefficients. This problem was solved before the end of the 19th century by various methods, among which the most constructive turned out to be the Routh–Hurwitz method.

In the transition to higher-order mathematical models, whose coefficients depend on several parameters (both deterministic and random) there arose a problem similar to Maxwell's: to estimate the signs of the real parts of the eigenvalues of matrices of general form depending on a vector-valued parameter, using only the entries of the matrix itself.

If for algebraic equations an explicit solution of the problem has been found in the form of determinantal inequalities [1], for matrices and matrix-valued functions of a matrix or vector argument the most general solution is still a theorem of Lyapunov's second method (1892), stating conditions for the spectrum of a matrix to be located in a certain domain of the complex plane depending on the existence of a solution of fixed sign of Lyapunov's matrix equation. In this connection the methods of perturbation

1980 *Mathematics Subject Classification* (1985 *Revision*). Primary 65F15, 34D20, 65H10; Secondary 15A52, 65H15.

Translation of Numerical-Analytical Methods for Investigating the Dynamics and Stability of Complex Systems (I. A. Lukovskiĭ, editor), Inst. Mat. Akad. Nauk Ukrain. SSR, Kiev, 1984, pp. 102–110; MR **87d**:65040.

theory were widely used to obtain an approximate dependence of the spectrum of a matrix-valued function on the parameters.

In the first stage the perturbation method for the algebraic eigenvalue problem was used in numerical analysis. In 1846 Jacobi obtained a first approximation formula for an eigenvalue of a symmetric perturbed matrix arising in the problem of deviations in the orbits of planets. This formula was subsequently generalized to matrices of general form, and has been included in many works on numerical methods of linear algebra (see [2]–[4]). In the next stage the perturbation method for estimating the eigenvalues of a matrix was used in the theory of linear differential equations with periodic coefficients (see, for instance, [5], Chapter IV, §1). Recently this approach has been used in stability theory and in control theory (see [6]–[8]). Its extension to the case of selfadjoint operators (first and foremost, in quantum field theory) led to the creation of a well-developed perturbation theory for linear operators (see [9]–[11]).

As we know, the essence of the perturbation method consists in expanding the function sought in a series of decreasing quantities. The expansion is carried out in a neighborhood of the point where the value of the function is given. In the parametric eigenvalue problem one uses formal series and Taylor series expansions of eigenvalues and eigenvectors. However, the high order of the matrices and the nonlinearity of the problem force nearly all authors to restrict themselves to find a first-order (sometimes second-order) approximation; higher-order approximations are usually obtained where the terms of the series are expressed in terms of matrices that need to be inverted (for instance, [5], Chapter IV, formula (1.40)). Moreover, the majority of the results are determined by the specific character of the original matrix and cannot be generalized even to similar cases. This is connected to a large extent with the coordinate form of the solution of the problem, when the series expansion is carried out separately for each eigenvalue and each eigenvector. The matrix form of the perturbation method in the eigenvalue problem suggested in [7] provided a universal approach to the solution of the problem and to the choice of the corresponding algorithm. This, in turn, allowed one to carry out a more complete analysis of the perturbation method as a whole and to obtain formulas that are suitable for a wide class of matrices depending on parameters. The generality of the matrix form of the solution allows one to choose an optimal algorithm for a concrete problem and to single out its essential objects. An important role here is played by a successfully chosen notation. We consider below the problems of constructing a general perturbation algorithm in the eigenvalue problem without inversion of series, and we also consider a priori estimates of convergence of the method as well as its specific features.

2. Let there be given an $n \times n$ matrix-valued function $A(p)$ of the m-dimensional parameter vector p:

$$A(p)\colon \mathbf{R}^m \to \mathbf{R}^n, \quad A(p) = A_0 + A_1(p), \qquad p \in P, \tag{1}$$

where $A_0 = A(p_0)$ and $A_1(p) = A(p) - A_0$; here p_0 is a given value of p, and P is a bounded domain of values of p.

We pose the problem of estimating the stability of the matrix (1), i.e., the problem of finding a criterion which would allow one to single out the values of the vector $p \in P$ for which the matrix $A(p)$ is stable. In what follows we consider the local case of the problem: we assume that the matrix A_0 is stable, and the domain P is sufficiently small. Among the various problems that require one to estimate the spectrum of a matrix-valued function $A(p)$ given its spectrum at a point, we mention the problem of studying the influence of manufacture tolerances on the stability and quality of motion of various controlled mechanical systems. For such systems one knows the rated (nominal) value of the parameter vector p_0 and the admissible limit deviations of these parameters from the nominal value (the domain P). In other words, the actual values of the parameters of the physical system belong to a given domain of the parameter space, but the precise point is unknown. Since the technological process tries to minimize the deviation of the parameter values, these deviations are quite small, and the perturbation method can be used to estimate their influence on the spectrum of the matrix. A second class of problems is connected with the design of control systems for which the vector p represents the controlled parameters. In this case local problems (small deviations of the parameters are necessary to obtain the required control properties) usually occur in the final design stage, when rough values of the controlled parameters are already known.

3. In order to find a functional dependence of the spectrum of the matrix-valued function $A(p)$ on the parameter vector p we use the matrix equation of the algebraic eigenvalue problem

$$A(p)S(p) - S(p)\Lambda(p) = 0, \tag{2}$$

where $S(p)$ is an $n \times n$ matrix-valued function taking $A(p)$ to the canonical form $\Lambda(p)$. In the sequel we assume that $\Lambda(p)$ is a diagonal matrix with distinct entries and the columns of $S(p)$ are eigenvectors of $A(p)$.

The matrix equation (2) is equivalent to a system of n^2 nonlinear homogeneous algebraic equations with $n(n+1)$ unknowns. Even for a fixed value of the parameter vector p the solution of (2) is one of the most complex problems of applied and pure mathematics. The existing numerical methods of solution of the eigenvalue problem may be used in the parametric case to find an initial approximation of the corresponding approximate method, and to estimate the structure of the matrix-valued function $A(p)$ at various points of the parameter domain P. The construction of the explicit solution of (2) is clearly possible only locally, in a neighborhood of some point of P where the solution has been found numerically. For nonsmall domains P one has to partition into subdomains and use the method of analytic continuation of series obtained by the methods of perturbation theory.

The solvability of (2) is guaranteed by the theorem of linear algebra on the canonical form of an arbitrary matrix, and by a sufficiently smooth dependence of the matrix on the parameter.

The traditional form of the perturbation method in the parametric eigenvalue problem consists in the passage from the matrix equation (2) to n systems of equations for each eigenvector and eigenvalue (see [2]–[6]). The matrix form of the perturbation method (see [7] and [8]), along with concise calculations and universal constructions, facilitates considerably the choice of the most acceptable approximate method of solution of a concrete problem, which is far more difficult to discover in the vector form.

4. Let us construct the solution (2) in matrix form for the case when p_0 is a point in general position for the matrix-valued function $A(p)$, i.e., the character of the spectrum of $A(p)$ is unchanged at the point itself and in a neighborhood of the point, and the spectrum has no multiplicities. We shall characterize the separation of the points of the spectrum for the matrix $A_0 = A(p_0)$ by the number $\delta = \min_{j,k} |\lambda_j^0 - \lambda_k^0|$ $(j, k = 1, \dots, n)$, where the λ_j^0 are the eigenvalues of A_0. We determine the smallness of the perturbation $A_1(p)$ by the relation $\|A_1(p)\| < \varepsilon \ll 1$ for all $p \in P$ ($\|\cdot\|$ is a matrix norm).

Let us first transform (2) to a form that takes into consideration the characteristic features of the method of successive approximations. To this end, we substitute the expression (1) for $A(p)$ into (2) and multiply the equation from the left by the nonsingular matrix $T_0 = S_0^{-1}$. We obtain an equation featuring the diagonal matrix $\Lambda_0 = T_0 A_0 S_0$ with the perturbation

$$[\Lambda_0 + B(p)]Z(p) - Z(p)\Lambda(p) = 0, \tag{3}$$

where $B(p) = T_0 A_1(p) S_0$ and $Z(p) = T_0 S(p)$. The solution (Λ_0, S_0) of (2) becomes the solution (Λ_0, Z_0) of (3). By the definition of $Z(p)$ in (3), we have $Z_0 = T_0 S_0 = E$ (E is the identity matrix).

In a small neighborhood of the solution (Λ_0, E) of (3), since $B(p)$ is small, $Z(p)$ is close to the identity matrix, and $\Lambda(p)$ is close to Λ_0, and therefore they may be represented as

$$\Lambda(p) = \Lambda_0 + \check{\Lambda}(p), \qquad Z(p) = E + \check{Z}(p), \tag{4}$$

where $\check{\Lambda}(p) = \Lambda(p) - \Lambda_0$ and $\check{Z}(p) = Z(p) - E$.

After substituting (4) in (3) and making simple rearrangements we obtain the equation for the corrections $\check{\Lambda}(p)$ and $\check{Z}(p)$ ("equation in variations"):

$$\check{\Lambda}(p) + \check{Z}(p) = \theta \circ [B(p) + B(p)\check{Z}(p) - \check{Z}(p)\Lambda(p)], \tag{5}$$

where θ is an $n \times n$ matrix with entries of the form $\theta_{jj} = 1$ and $\theta_{jk} = 1/(\lambda_k^0 - \lambda_j^0)$, $j, k = 1, \dots, n$, and $V = \theta \circ U$ is the Schur matrix product of θ and $U = (u_{jk})$, $v_{jk} = \theta_{jk} \cdot u_{jk}$.

Let us write the solution $(\check{\Lambda}(p), \check{Z}(p))$ of (5) as the formal series with decreasing terms

$$\check{\Lambda}(p) = \Lambda_1(p) + \Lambda_2(p) + \cdots, \qquad \check{Z}(p) = Z_1(p) + Z_2(p) + \cdots.$$

Taking into account the order of the terms in the series for $\check{\Lambda}(p)$ and $\check{Z}(p)$, we write down in the standard way from (5) the equation for the kth approximation $(\Lambda_k(p), Z_k(p))$:

$$\Lambda_k(p) + Z_k(p) = \theta_0 \left[B(p)Z_{k-1}(p) - \sum_{m=1}^{k-1} Z_m(p)\Lambda_{k-m}(p) \right]. \tag{6}$$

The infinite system of linear recurrent matrix equations (6) is equivalent to the original equations (2) and (5) and is an explicit multi-step iterative process for the approximate construction of the solutions $(\Lambda(p), S(p))$ and $(\Lambda(p), Z(p))$ respectively.

Equations (5) and (6) may be substantially simplified if we consider their degree of indetermination and the specific features of the matrices $\check{\Lambda}(p)$ and $\check{Z}(p)$. In fact, setting n diagonal entries of $\check{Z}(p)$ equal to zero, we obtain specific equations and guarantee their solvability (since $\Lambda(p)$ is diagonal). In this case (5) and (6) turn into the system of two matrix equations:

$$\check{\Lambda}(p) = F^d(p), \qquad \check{Z}(p) = F^0(p),$$

where $F(p) = \theta \circ [B(p) + B(p)\check{Z}(p) - \check{Z}(p)\check{\Lambda}(p)]$; $F^0(p)$ and $F^d(p)$ are matrices obtained from $F(p)$ by equating to zero respectively the diagonal and the nondiagonal entries;

$$\Lambda_k(p) = F_k^d(p), \qquad Z_k(p) = F_k^0(p), \tag{7}$$

where $F_k(p) = \theta \circ [B(p)Z_{k-1}(p) - \sum_{m=1}^{k-1} Z_m(p)\Lambda_{k-m}(p)]$.

The analysis of the iterative procedure given by the explicit expressions (6) and (7) for $\Lambda_k(p)$ and $Z_k(p)$ shows that the construction of each approximation requires the realization of three basic operations: construction of the matrix $[B(p)Z_{k-1}(p) - \sum_{m=1}^{k-1} Z_m(p)\Lambda_{k-m}(p)]$, division of the jkth entry of this matrix by $\lambda_k^0 - \lambda_j^0$, and construction of the matrices $F_k^d(p)$ and $F_k^0(p)$. Clearly, for the diagonal matrix $\Lambda(p)$ the contribution in the matrix $F_k^d(p)$ $(\Lambda_k(p))$ comes only from the product $\theta \circ B(p)Z_{k-1}(p)$, and the sum $\theta \circ \sum_{m=1}^{k-1} Z_m(p)\Lambda_{k-m}(p)$ has zero diagonal. Consequently, the kth term of the series expansion of the solution of (5) ((2)) is determined by the initial and the $k-1$ previous terms, and the kth term of the series for $\check{\Lambda}(p)$ is determined by the $(k-1)$th term of the series for $\check{Z}(p)$. Hence, despite the explicit representation (7) of the kth terms, the multi-step character of the iterative process admits only qualitative a priori estimates of the convergence of the corresponding series. For the assumed properties of the matrix A_0 the convergence of the series for $\Lambda(p)$ and $Z(p)$ is determined by the degree of smallness of the quotient ε/δ.

5. Let us use (6) and (7) to write down expressions for the entries of the matrices $\Lambda_k(p) = (\lambda_i^{(k)})$ and $Z_k(p) = (z_{ij}^{(k)})$ (for simplicity in the notation we omit the dependence on p):

$$z_{ij}^{(k)} = \frac{1}{\lambda_j^0 - \lambda_i^0}\left[\sum b_{ir} z_{rj}^{(k-1)} - \sum z_{ij}^{(m)} \lambda_j^{(k-m)}\right], \qquad i \neq j,\ z_{ij}^{(k)} = 0,$$
$$\lambda_i^{(k)} = \sum_{r=1}^{n} b_{ir} z_{ri}^{(k-1)}, \qquad k = 1, 2, \dots;\ 1 \leq i, j \leq n. \tag{8}$$

From (8) it is not difficult to obtain expressions for the first terms of the series for $\check{\Lambda}(p)$ and $\check{Z}(p)$ in terms of the spectrum of A_0 and the perturbation $A_1(p)$ (here $i, j = 1, \dots, n$ and $i \neq j$):

$$\lambda_i^{(1)} = b_{ii}, \qquad z_{ii}^{(1)} = 0;$$
$$z_{ij}^{(1)} = b_{ij}/(\lambda_j^0 - \lambda_i^0);$$
$$\lambda_i^{(2)} = \sum_{r=1}^{n} \frac{b_{ir} b_{ri}}{\lambda_i^0 - \lambda_r^0};$$
$$z_{ij}^{(2)} = \frac{1}{\lambda_j^0 - \lambda_i^0}\left[\sum_{r=1}^{n} \frac{b_{ir} b_{rj}}{\lambda_j^0 - \lambda_r^0} - \frac{b_{ij} b_{jj}}{\lambda_j^0 - \lambda_i^0}\right]; \qquad z_{ii}^{(2)} = 0; \tag{9}$$
$$\lambda_i^{(3)} = \sum_{\substack{r=1\\ r\neq i}}^{n} \frac{b_{ir}}{\lambda_i^0 - \lambda_r^0}\left[\sum_{\substack{s=1\\ s\neq i}}^{n} \frac{b_{rs} b_{si}}{\lambda_i^0 - \lambda_s^0} - \frac{b_{ri} b_{ii}}{\lambda_i^0 - \lambda_r^0}\right];$$
$$z_{ij}^{(3)} = \frac{1}{\lambda_j^0 - \lambda_i^0}\left\{\sum_{\substack{r=1\\ r\neq j}}^{n} \frac{b_{ir}}{\lambda_j^0 - \lambda_r^0}\left[\sum_{\substack{s=1\\ s\neq j}}^{n} \frac{b_{is} b_{sj}}{\lambda_j^0 - \lambda_s^0} - \frac{b_{rj} b_{jj}}{\lambda_j^0 - \lambda_r^0}\right]\right.$$
$$\left. - \frac{b_{ij}}{\lambda_j^0 - \lambda_i^0} \sum_{\substack{r=1\\ r\neq j}}^{n} \frac{b_{jr} b_{rj}}{\lambda_j^0 - \lambda_r^0} - \frac{b_{jj}}{\lambda_j^0 - \lambda_i^0}\left[\sum_{\substack{r=1\\ r\neq j}}^{n} \frac{b_{ir} b_{rj}}{\lambda_j^0 - \lambda_r^0} - \frac{b_{ij} b_{jj}}{\lambda_j^0 - \lambda_i^0}\right]\right\};$$

and $z_{ii}^{(3)} = 0$. The entries of $B(p)$ figuring in (9) may be expressed in terms of the entries of $A_1(p)$, T_0, and S_0 as the bilinear form

$$b_{ij} = t_i A_i(p) s_j, \qquad i, j = 1, \dots, n,$$

where t_i is the ith row of T_0 and s_j is the jth column of S_0.

6. As an illustrative example, let us consider the case $n = 2$:

$$A_0 = \begin{pmatrix} 0 & 1 \\ -\omega^2 & -2h \end{pmatrix}, \qquad A_1(p) = \begin{pmatrix} 0 & 0 \\ a_1(p) & a_2(p) \end{pmatrix},$$
$$\Lambda_0 = \begin{pmatrix} \lambda_1^0 & 0 \\ 0 & \lambda_2^0 \end{pmatrix}, \qquad S_0 = \begin{pmatrix} 1 & 1 \\ \lambda_1^0 & \lambda_2^0 \end{pmatrix}, \qquad T_0 = \gamma \begin{pmatrix} \lambda_2^0 & -1 \\ -\lambda_1^0 & 1 \end{pmatrix},$$

where $\lambda_{1,2}^0 = -h \pm \mathrm{id}$, $d^2 = \omega^2 - h^2$, $\omega^2 \geq h^2$, $i = \sqrt{-1}$, $\delta = 2d$, and $\gamma = \frac{i}{2d}$.

We find the matrix $B(p) = T_0 A_1(p) S_0$:

$$B(p) = \gamma \begin{pmatrix} -b_1 & -b_2 \\ b_1 & b_2 \end{pmatrix},$$

where $b_1 = a_1(p) + \lambda_1^0 a_2(p)$ and $b_2 = a_1(p) + \lambda_2^0 a_2(p)$.

Using either (7) or (8) and (9), we find the first four terms of the series for $\check{\Lambda}(p)$ and $\check{Z}(p)$:

$$\Lambda_1(p) = \gamma \begin{pmatrix} -b_1 & 0 \\ 0 & b_2 \end{pmatrix},$$

$$Z_1(p) = \gamma^2 \begin{pmatrix} 0 & -b_2 \\ -b_1 & 0 \end{pmatrix},$$

$$\Lambda_2(p) = \gamma^3 b_1 b_2 \begin{pmatrix} 1 & 0 \\ 0 & -1 \end{pmatrix},$$

$$Z_2(p) = \gamma^4 (b_1 + b_2) \begin{pmatrix} 0 & b_2 \\ b_1 & 0 \end{pmatrix},$$

$$\Lambda_3(p) = \gamma^5 b_1 b_2 (b_1 + b_2) \begin{pmatrix} -1 & 0 \\ 0 & 1 \end{pmatrix},$$

$$Z_3(p) = \gamma^6 [b_1 b_2 + (b_1 + b_2)^2] \begin{pmatrix} 0 & -b_2 \\ -b_1 & 0 \end{pmatrix},$$

$$\Lambda_4(p) = \gamma^7 b_1 b_2 [b_1 b_2 + (b_1 + b_2)^2] \begin{pmatrix} 1 & 0 \\ 0 & -1 \end{pmatrix},$$

$$Z_4(p) = \gamma^8 (b_1 + b_2)[3b_1 b_2 + (b_1 + b_2)^2] \begin{pmatrix} 0 & b_2 \\ b_1 & 0 \end{pmatrix}.$$

Then the matrix $\Lambda(p)$ has the following form, up to fifth-order terms:

$$\Lambda(p) = \begin{pmatrix} \lambda_1(p) & 0 \\ 0 & \lambda_2(p) \end{pmatrix},$$

$$\lambda_{1,2}(p) \cong -h + \frac{a_2(p)}{2} \pm i \left(d - \frac{\alpha}{2d} - \frac{\alpha^2 + \beta}{8d^3} - \frac{\alpha^3 + \alpha\beta}{16d^5} - \frac{5\alpha^4 + b\alpha^2\beta + \beta^2}{128d^7} - \cdots \right),$$

where $\alpha = a_1(p) - h a_2(p)$ and $\beta = d^2 a_2^2(p)$.

This agrees completely with the corresponding Taylor series expansion of the variables $\lambda_{1,2}(p)$. Here the series converge if $\alpha/\delta < 1$.

7. Our suggested approach to the solution of equation (2) of the parametric eigenvalue problem allows us to find explicitly the dependence of the real parts of the eigenvalues of $A(p)$ directly in terms of its entries, i.e., to construct a stability criterion for $A(p)$ (and the corresponding differential equation), and thus to solve Maxwell's problem. Starting from the computed expressions for the spectrum of the matrix $A(p)$, one can also then carry out an estimate of the (asymptotic) stability domain, or choose a suitable value of the parameter p for which the matrix $A(p)$ is stable and has additional properties (the synthesis problem). If the parameter domain P is sufficiently small and the spectrum of A_0 satisfies the conditions indicated above, then in concrete computations it suffices to take the first two terms of the expansion for $\check{\Lambda}(p)$, using subsequent terms of the series to estimate how good the approximation is.

BIBLIOGRAPHY

1. N. G. Chebotarev and N. N. Meĭman, *The Routh–Hurwitz problem for polynomials and entire functions*, Trudy Mat. Inst. Steklov. **26** (1949). (Russian)

2. D. K. Faddeev and V. N. Faddeeva, *Computational methods of linear algebra*, 2nd ed., Fizmatgiz, Moscow, 1963; English transl. of 1st ed., Freeman, San Francisco, Calif., 1963.

3. V. V. Voevodin, *Numerical methods of algebra. Theory and algorithms*, "Nauka", Moscow, 1966. (Russian)

4. J. H. Wilkinson, *The algebraic eigenvalue problem*, Clarendon Press, Oxford, 1965.

5. V. A. Yakubovich and V. M. Starzhinskiĭ, *Linear differential equations with periodic coefficients and their applications*, "Nauka", Moscow, 1972; English transl., Vols. I, II, Israel Program Sci. Transls., Jerusalem, and Wiley, New York, 1975.

6. K. A. Abgaryan, *Matrix and asymptotic methods in the theory of linear systems*, "Nauka", Moscow, 1973. (Russian)

7. N. A. Pustovoĭtov, *Perturbation method in the algebraic eigenvalue problem*, Preprint No. 76. 31, Inst. Mat. Akad. Nauk Ukrain. SSR, Kiev, 1976. (Russian) MR **58** #31783.

8. A. M. Mikhaĭlichenko, N. A. Pustovoĭtov, and V.G. Sukhorebriĭ, *Stability of control systems with random parameters*, "Naukova Dumka", Kiev, 1981. (Russian)

9. Kurt O. Friedrichs, *Perturbation of spectra in Hilbert space*, Amer. Math. Soc., Providence, R.I., 1965.

10. Nelson Dunford and Jacob T. Schwartz, *Linear operators*. Vol. 1: *General theory*, Interscience, 1958.

11. Tosio Kato, *Perturbation theory for linear operators*, Springer-Verlag, 1966.

Translated by J. J. TOLOSA

Amer. Math. Soc. Transl.
(2) Vol. **144**, 1989

Eigenvalues and Eigenfunctions of the Peierls Equation

I. V. LOBAREV

1. We will consider the eigenvalue problem for the transport equation

$$\mu\frac{\partial\varphi}{\partial x}+\varphi(x,\mu)=\frac{\lambda}{2}\int_{-1}^{1}\varphi(x,\mu)\,d\mu \quad (x\in[-1,1],\ \mu\in[-1,1]) \tag{1}$$

with the boundary conditions

$$\begin{aligned}\varphi(1,\mu)&=0, &\quad \mu<0,\\ \varphi(-1,\mu)&=0, &\quad \mu>0.\end{aligned} \tag{2}$$

Problem (1), (2) is equivalent to the solution of a system of the two equations [1]:

$$L\varphi=\lambda\Phi(x), \tag{3}$$

$$\Phi=\lambda K\Phi, \tag{4}$$

where

$$K\Phi=\frac{1}{2}\int_{-1}^{1}E_1(|x-x'|)\Phi(x')\,dx',$$

$$L\varphi=\mu\frac{\partial\varphi}{\partial x}+\varphi(x,\mu),$$

$$\Phi(x)=\frac{1}{2}\int_{-1}^{1}\varphi(x,\mu)\,d\mu,$$

$$E_1(x)=\int_1^\infty\frac{e^{-yx}}{y}\,dy.$$

If the solution of equation (4), i.e., λ and the corresponding eigenfunction $\Phi(x)$, is known, then, using invertibility in an explicit form of the operator L, one can construct a solution $\varphi(x,\mu)$ of equation (3): $\varphi(x,\mu)=\lambda L^{-1}\Phi(x)$. Conversely, if a solution of (3), i.e., λ and the corresponding eigenfunction

1980 *Mathematics Subject Classification* (1985 *Revision*). Primary 45K05, 45C05.

Translation of Sem. Methods Numer. Appl. Math. (G. I. Marchuk, director), Preprint No. 84-109, Vychisl. Tsentr Sibirsk. Otdel. Akad. Nauk SSSR, Novosibirsk, 1984; MR **87i**: 82113.

$\varphi(x,\mu)$, is known, then $\Phi(x) = \frac{1}{2}\int_{-1}^{1}\varphi(x,\mu)\,d\mu$ is an eigenfunction of equation (4), and the eigenvalue λ corresponds to it. Therefore, we will consider only the Peierls integral equation (4).

The eigenvalues of problem (4), arranged in order of increasing absolute value, will be denoted by λ_k, repeating each eigenvalue as many times as its multiplicity, and their corresponding eigenfunctions will be denoted by $\Phi_k(x)$. It is well known [1] that for our problem the set $\{\lambda_k\}$ of eigenvalues is countable, does not have finite limit points, and consists of positive numbers. The functions $\Phi_k(x) \in W_1[-1,1]$, where

$$\|f\|_{W_1} = \|f\|_C + \sup_{x,x'} \frac{|f(x)-f(x')|}{\omega(|x-x'|)} < \infty,$$
$$\omega(y) = cy(1+|\ln y|), \qquad y>0,\ c=\text{const},$$

can be chosen to satisfy the condition $(\Phi_l,\Phi_m)_{L_2[-1,1]} = \delta_{l,m}$ (here $\delta_{l,m}$ is the Kronecker symbol). The smallest eigenvalue λ_0 is simple, and the corresponding function $\Phi_0(x)$ is strictly positive on $[-1,1]$. The orthonormal system $\{\Phi_l\}_0^\infty$ is complete in $L_2[-1,1]$. It has also been proved that the operator K is symmetric, positive definite, and compact in $L_2[-1,1]$. Therefore the invariant subspace corresponding to $\lambda=\lambda_k$ coincides with the eigensubspace, and the dimension of the eigensubspace (multiplicity of the eigenvalue) is finite.

In this paper we show that all the eigenvalues are simple, i.e., that each has multiplicity one, and we give two-sided estimates for all the eigenvalues.

LEMMA 1. *All the eigenfunctions $\Phi_i(x)$, $i=1,\dots,n$, of problem* (4) *corresponding to the eigenvalue λ can be chosen to form an orthonormal system such that each of these functions has parity* (*i.e., it is either an even or an odd function*).

PROOF. The eigenfunctions $\Phi_i(x)$ $(i=1,\dots,n)$ corresponding to the eigenvalue λ generate an eigensubspace E_λ of dimension n, and the functions $\Phi_i(x)$ $(i=1,\dots,n)$ form a basis for E_λ. The change of variables $t=-x$, $t'=-x'$ in (4) implies that the system of functions $F_i(x)=\Phi_i(-x)$ forms a basis for E_λ. Obviously, any linear combination of functions of the system $\{\Phi_i(x),\ F_i(x)\}_1^n$ does not lead us out of E_λ, and so lies in E_λ.

Successively, we construct two orthonormal systems of functions (even $U_i^+(x)$, $i=1,\dots,n$, and odd $U_i^-(x)$, $i=1,\dots,n$) as linear combinations of the system $\{\Phi_i(x),\ F_i(x)\}_1^n$. To construct a system of even functions we apply the successive method of orthogonalization:

$$U_1^+(x) = \frac{1}{H_1}(\Phi_1(x) + F_1(x)), \qquad H_1 = \|\Phi_1(x) + F_1(x)\|_{L_2[-1,1]},$$

$$U_2^+(x) = \frac{1}{H_2}(\Phi_2(x) + F_2(x) - c_2^1 U_1^+(x)),$$

$$H_2 = \|\Phi_2(x) + F_2(x) - c_2^1 U_1^+(x)\|_{L_2^1[-1,1]},$$

$$C_2^1 = (\Phi_2(x) + F_2(x), U_1^+(x))_{L_2[-1,1]},$$

....

$$C_n^k = (\Phi_n(x) + F_n(x), U_k^+(x))_{L[-1,1]}, \qquad k = 1, \dots, n-1,$$

$$U_n^+(x) = \frac{1}{H_n}\left(\Phi_n(x) + F_n(x) - \sum_{k=1}^{n-1} C_n^k U_k^+(x)\right),$$

$$H_n = \left\|\Phi_n(x) + F_n(x) - \sum_{k=1}^{n-1} C_n^k U_k^+(x)\right\|_{L[-1,1]}.$$

We remark that if for some l

$$\Phi_l(x) + F_l(x) - \sum_{k=1}^{l-1} E_l^k U_k^+(x) \equiv 0,$$

then we set $U_l^+(x) = 0$.

The analogous process of orthogonalization leads to a system of odd functions $\{U_i^-(x)\}_1^n$ when in the above expressions $F_i(x)$ is replaced by $-F_i(x)$. Since $U_i^+(x)$ is an even function and $U_j^-(x)$ is an odd function, we have

$$(U_i^+(x), U_j^-(x))_{L_2[-1,1]} = 0$$

for all $i, j = 1, \dots, n$. Thus, the system of functions

$$\{U_1^+(x), \dots, U_n^+(x), U_1^-(x), \dots, U_n^-(x)\} \tag{5}$$

is orthonormal and belongs to E_λ. Since the dimension of E_λ is n, the number of nontrivial functions in the system (5) does not exceed n. We will show that it is n.

Suppose this is not true; then there exists a nontrivial eigenfunction $\Phi(x) \in E_\lambda$ which satisfies the equalities

$$(\Phi(x), U_k^+(x)) = 0, \qquad k = 1, \dots, n, \tag{6}$$

$$(\Phi(x), U_k^-(x)) = 0, \qquad k = 1, \dots, n. \tag{7}$$

The definition of system (5) and equalities (6) and (7) successively imply that for all $k = 1, \dots, n$

$$(\Phi(x), \Phi_k(x))_{L_2[-1,1]} = 0.$$

This successive process can be traced along the following chain of equalities:

$$\left.\begin{array}{l}(\Phi_1(x)+F_1(x),\Phi(x))_{L_2[-1,1]}=0\\(\Phi_1(x)-F_1(x),\Phi(x))_{L_2[-1,1]}=0\end{array}\right\}$$

$$\Rightarrow \left.\begin{array}{l}(\Phi(x),\Phi_1(x))_{L_2[-1,1]}=(\Phi(x),F_1(x))_{L_2[-1,1]}\\(\Phi(x),\Phi_1(x))_{L_2[-1,1]}=-(\Phi(x),F_1(x))_{L_2[-1,1]}\end{array}\right\}$$

$$\Rightarrow \left.\begin{array}{l}(\Phi(x),\Phi_1(x))_{L_2[-1,1]}=0\\(\Phi(x),F_1(x))_{L[-1,1]}=0\end{array}\right\}$$

$$\Rightarrow \left.\begin{array}{l}(\Phi(x),\Phi_2(x)+F_2(x))_{L_2[-1,1]}=0\\(\Phi(x),\Phi_2(x)-F_2(x))_{L_2[-1,1]}=0\end{array}\right\}.$$

Thus we have a contradiction: the nonzero function $\Phi(x)\in E_\lambda$ is orthogonal to the basis $\Phi_1(x),\dots,\Phi_n(x)$ of E_λ. By the same token we have shown that when the trivial functions are removed from the orthonormal system of functions (5), we obtain an orthonormal basis for the subspace E_λ, which has parity. The lemma is proved.

2. We will consider the following integral equation with continuous kernel which depends on the parameter $t\in(0,\infty)$:

$$\mu_k(t)U_k^t(x)=\int_{-1}^{1}e^{-t|x-x'|}U_k^t(x')\,dx'. \tag{8}$$

The eigenfunctions of equation (8) are continuous. We will consider solutions of (8) in the Hilbert space $L_2[-1,1]$.

LEMMA 2. *The functions*

$$U_{2k}^t(x)=\cos f_{2k}(t)x,\qquad k=0,1,\dots,$$

are the eigenfunctions of equation (8). *Their corresponding eigenvalues are*

$$\mu_{2k}(t)=\frac{2t}{t^2+f_{2k}^2(t)},\qquad k=0,1,\dots.$$

Here the function $\tau=f_{2k}(t)$ maps $[0,\infty)$ into $[\pi k,\pi k+\frac{\pi}{2})$ and is the kth branch of a multi-valued function $F^{-1}(t)$, where $F(\tau)=\tau\tan\tau$.

PROOF. First we remark that since the multi-valued function $F^{-1}(\tau)$ has a countable number of branches, analogously to the function $\operatorname{Arctan}\tau$, we obtain a countable set of functions $\{f_{2k}(t)\}_{k=0}^{\infty}$ and the same number of solutions for our problem.

We remark that on each interval $[\pi k,\pi k+\frac{\pi}{2})$ the function $F(\tau)$ is monotonically increasing; therefore $f_{2k}(t)$, which maps $[\pi k,\pi k+\frac{\pi}{2})$ into $[0,\infty)$, is also monotonically increasing.

We proceed to the proof of the lemma. We substitute $u_{2k}^t(x) = \cos(f_{2k}(t)x)$ into the right-hand side of (8), which results in the following chain of equalities:

$$\begin{aligned}
&\int_{-1}^{1} e^{-t|x-x'|}\cos(f_{2k}(t)x')\,dx' \\
&\quad = \int_{-1}^{x} e^{-tx}(e^{tx'}\cos f_{2k}(t)x')\,dx' + e^{tx}\int_{x}^{1} e^{-tx'}\cos f_{2k}(t)x'\,dx' \\
&\quad = e^{-tx}\left(\frac{e^{tx'}}{t^2+f_{2k}^2(t)}(t\cos f_{2k}(t)x' + f_{2k}(t)\sin f_{2k}(t)x')|_{-1}^{x}\right) \\
&\qquad + e^{tx}\left(\frac{e^{-tx'}}{t^2+f_{2k}^2(t)}(-t\cos f_{2k}(t)x' + f_{2k}(t)\sin f_{2k}(t)x')|_{x}^{1}\right) \\
&\quad = \frac{2t}{t^2+f_{2k}^2(t)}\cos f_{2k}(t)x - 2\cosh tx\,\frac{e^{-t}(t\cos f_{2k}(t) - f_{2k}(t)\sin f_{2k}(t))}{t^2+f_{2k}^2(t)}.
\end{aligned}$$

Taking into account that $\tau = f_{2k}(t)$, where $t = F(\tau)$ $(\tau \in [\pi k, \pi k + \frac{\pi}{2}))$, we obtain

$$F(\tau)\cos\tau - \tau\sin\tau = \tau\tan\tau\cos\tau - \tau\sin\tau \equiv 0.$$

The lemma is proved.

LEMMA 3. *The functions*

$$U_{2k+1}^t(x) = \sin\psi_{2k+1}(t)x, \qquad k = 0, 1, \ldots,$$

are eigenfunctions of equation (5). *Their corresponding eigenvalues are*

$$\mu_{2k+1}(t) = \frac{2t}{t^2+\psi_{2k+1}^2(t)}, \qquad k = 0, 1, \ldots.$$

Here the function $\tau = \psi_{2k+1}(t)$ *maps* $[0,\infty)$ *into* $[\pi k + \frac{\pi}{2}, \pi(k+1))$ *and is the* k*th branch of a multi-valued function* $\Psi^{-1}(t)$, *where* $\Psi(\tau) = -\tau\cot\tau$.

PROOF. Everything noted in the previous lemma also pertains to the proof of this lemma, and the function $\psi_{2k+1}(t)$, which maps $[\pi k + \frac{\pi}{2}, \pi(k+1))$ into $[0,\infty)$, is monotonically increasing. As in the previous lemma, we substitute

$U^t_{2k+1}(x) = \sin \psi_{2k+1}(t)x$ into the right-hand side of (8). This yields

$$\begin{aligned}
&\int_{-1}^{1} e^{-t|x-x'|} \sin(\psi_{2k+1}(t)x')\,dx' \\
&= e^{-tx} \int_{-1}^{x} \sin \psi_{2k+1}(t)x'\,dx' + e^{tx} \int_{x}^{1} e^{-tx'} \sin \psi_{2k+1}(t)x'\,dx' \\
&= e^{-tx} \left(\frac{e^{tx'}}{t^2 + \psi^2_{2k+1}(t)} (t \sin \psi_{2k+1}(t)x' - \psi_{2k+1}(t) \cos \psi_{2k+1}(t)x' |_{-1}^{x} \right) \\
&\quad + e^{tx} \left(\frac{e^{-tx'}}{t^2 + \psi_{2k+1}(t)} (-t \sin \psi_{2k+1}(t)x' - \psi_{2k+1}(t) \cos \psi_{2k+1}(t)x')|_{x}^{1} \right) \\
&= \frac{2t}{t^2 + \psi^2_{2k+1}(t)} \sin \psi_{2k+1}(t)x \\
&\quad - 2 \sinh tx \frac{e^{-t}(t \sin \psi_{2k+1}(t) + \psi_{2k+1}(t) \cos \psi_{2k+1}(t))}{t^2 + \psi^2_{2k+1}(t)}.
\end{aligned}$$

Taking into account that $\tau = \psi_{2k+1}(t)$, where $t = \Psi(\tau)$ $(\tau \in [\frac{\pi}{2} + k\pi, (k+1)\pi))$, we get

$$\Psi(\tau) \sin \tau + \tau \cos \tau = -\tau \cot \tau \sin \tau + \tau \cos \tau \equiv 0.$$

Thus, Lemma 3 is proved.

THEOREM 1. *The integral equation* (8) *has eigenfunctions*

$$U^t_{2k}(x) = \cos(f_{2k}(t)x), \tag{9}$$

$$U^t_{2k+1}(x) = \sin(\psi_{2k+1}(t)x) \tag{10}$$

and their corresponding eigenvalues are

$$\mu_{2k}(t) = \frac{2t}{t^2 + f^2_{2k}(t)}, \tag{11}$$

$$\mu_{2k+1}(t) = \frac{2t}{t^2 + \psi^2_{2k+1}(t)}, \tag{12}$$

where $f_{2k}(t)$ *and* $\psi_{2k+1}(t)$ *are defined in Lemmas* 2 *and* 3. *Equation* (8) *does not have any other eigenfunctions in* $L[-1, 1]$.

PROOF. The fact that (9) and (10) are eigenfunctions follows from Lemmas 2 and 3. The completeness of the system of eigenfunctions (9), (10) in $L_2[-1, 1]$ can be proved in different ways. Our proof will use the apparatus of oscillation kernels [3]. Since for each fixed $t \neq 0$ the operator in (8) is compact, and therefore the system of eigenfunctions is complete in $L_2[-1, 1]$, it remains to show that the operator in (8) does not have eigenfunctions other than the functions (9) and (10).

For this purpose we will introduce some definitions, notation, and statements from [3].

If $K(x,s)$ is a continuous kernel defined on $[-a,a]\times[-a,a]$, we introduce the notation

$$K\begin{pmatrix} x_1,\dots,x_n \\ s_1,\dots,s_n \end{pmatrix} = \begin{vmatrix} K(x_1,s_1) & \dots & K(x_1,s_n) \\ \vdots & \ddots & \vdots \\ K(x_n,s_1) & \dots & K(x_n,s_n) \end{vmatrix},$$

where n is an arbitrary natural number, and $x_1,\dots,x_n$ and $s_1,\dots,s_n$ are arbitrary points from $[-a,a]$.

DEFINITION 1. A kernel $K(x,s)$, $-a \le s,x \le a$, is called an *oscillation kernel* if the following three conditions hold:

1. $K(x,s) > 0$ if $x,s \in [-a,a]$ with $\{x,s\} \ne \{-a,a\}$.
2. $K\binom{x_1,\dots,x_n}{s_1,\dots,s_n} \ge 0$ if $-a < x_1 < \dots < x_n < a$ and $-a < s_1 < \dots < s_n < a$ $(n = 1,2,\dots)$.
3. $K\binom{x_1,\dots,x_n}{x_1,\dots,x_n} > 0$ if $-a < x_1 < \dots < x_n < a$ $(n = 1,2,\dots)$.

DEFINITION 2. Let $\varphi(x) \in C(-a,a)$ be a function which vanishes at some interior point c_0. Then c_0 is called a *node* of the function $\varphi(x)$ if in any neighborhood of c_0 on opposite sides of c_0 one can find points x_1 and x_2 $(x_1 < c_0 < x_2)$ such that $\varphi(x_1)\varphi(x_2) < 0$.

ASSERTION 1 ([3], CHAPTER IV, §7.1). *A single-pair kernel*

$$K(x,s) = \begin{cases} \chi(x)\psi(s), & x \ge s, \\ \chi(s)\psi(x), & x \le s, \end{cases}$$

is an oscillation kernel if and only if $\chi(x)\psi(x) > 0$ *for* $-a < x < a$ *and the function* $\psi(x)/\chi(x)$ *is monotonically increasing.*

The following assertion is a particular case of Theorem 3 from [3] (Chapter IV, §3.7).

ASSERTION 2. *If a continuous symmetric kernel* $K(x,s)$, $-a < x,s < a$, *is an oscillation kernel, then all eigenvalues of the integral equation*

$$\mu\varphi(x) = \int_{-a}^{a} K(x,s)\varphi(s)\,ds$$

are simple. In addition, the following assertions are true:

A. $\varphi_0(x)$ *does not have zeros in* $(-a,a)$.

B. $\varphi_j(x)$ *has exactly* j *nodes in* $(-a,a)$ *and has no other zeros in* $(-a,a)$.

COROLLARY. *The kernel*

$$e^{-t|x-s|} = \begin{cases} e^{-tx}e^{ts}, & x \ge s, \\ e^{-ts}e^{tx}, & x \le s, \end{cases}$$

is an oscillation kernel for each fixed $t > 0$ *and therefore possesses properties* A *and* B.

Thus, in order to show that equation (8) does not have any other eigenfunctions, it is sufficient to show that $U_k^t(x)$ has k nodes. We will show that the function $U^t_{2k}(x) = \cos f_{2k}(t)x$, where $x \in [-1,1]$, has $2k$ nodes. For $U^t_{2k+1}(x) = \sin \psi_{2k+1}(t)x$ the proof is similar.

We make the change of variables $y = f_{2k}(t)x$. Now we have to calculate the number of nodes of the function $\cos y$ for $y \in [-f_{2k}(t), f_{2k}(t)]$. The definition of $f_{2k}(t)$ implies that for $t \in [0,\infty)$ the values of $f_{2k}(t)$ are in $[\pi k, \pi k + \frac{\pi}{2})$. The function $\cos y$ for $y \in [-f_{2k}(0), f_k(0)] = [-\pi k, \pi k]$ has $2k$ nodes. The remark in Lemma 2 implies that $f_{2k}(t)$ is monotone increasing; therefore, when t increases, the number of nodes can only increase. But $f_{2k}(t) \to \pi k + \frac{\pi}{2}$ as $t \to \infty$, and for $y \in [-\pi k - \frac{\pi}{2}, \pi k + \frac{\pi}{2}]$ the function $\cos y$ also has $2k$ nodes. Therefore, $U^t_{2k}(x) = \cos f_{2k}(t)x$ has exactly $2k$ nodes for all $t \in [0,\infty)$ and $x \in [-1,1]$. Similarly, the definition of $z = \psi_{2k+1}(t)$ implies that the function $U^t_{2k+1}(x) = \sin \psi_{2k+1}(t)x$ has $2k+1$ nodes.

Thus, the theorem is proved.

For each fixed t we obtain a complete (in $L_2[-1,1]$) orthonormal system $U^t_l(x)$. In what follows the norm of $U^t_l(x)$ will be denoted by $H_l(t)$.

We remark that for $t = 0$ we have $f_{2k}(0) = k\pi$ and $\psi_{2k+1}(0) = k\pi + \frac{\pi}{2}$, and the orthonormal system $U^0_{2k}(x) = \cos k\pi x$, $U^0_{2k+1}(x) = \sin(k+\frac{1}{2})\pi x$ is complete in $L_2[-1,1]$. As $t \to \infty$, $f_{2k}(t) \to (k+\frac{1}{2})\pi$, and $\psi_{2k+1}(t) \to (k+1)\pi$. The orthonormal system

$$U^\infty_{2k}(x) = \cos(k+\tfrac{1}{2})\pi x, \qquad U^\infty_{2k+1}(x) = \sin(k+1)\pi x$$

is also complete in $L_2[-1,1]$. Therefore, in what follows we will assume that $t \in [0,\infty)$.

3. Before continuing our arguments further, we prove a few auxiliary lemmas. The proof of the following assertion can be found, for example, in [1], [4], or [5].

ASSERTION 3. *For each $\varphi(x) \in L_2[-1,1]$ and for almost all $x \in [-1,1]$,*

$$\int_{-1}^{1}\left(\int_1^\infty \frac{e^{-t|x-x'|}}{t}\,dt\right)\varphi(x')\,dx' = \int_1^\infty\left(\int_{-1}^{1}\frac{e^{-t|x-x'|}}{t}\varphi(x')\,dx'\right)dt.$$

The proof follows directly from Fubini's theorem.

In what follows, we will reverse the order of integration without saying so.

LEMMA 4.

$$\int_{-1}^{1}\int_{-1}^{1} e^{-t|x-x'|}U^y_k(x)U^y_l(x')\,dx'\,dx$$
$$= \begin{cases} 2\left(\dfrac{tH^2_{2k}(y)}{t^2+f^2_{2k}(y)} - \dfrac{\cos^2 f_{2k}(y)}{(t+f^2_{2k}(y))^2}[(t^2-y^2)-e^{-2t}(t-y)^2]\right), & k = l;\\ -\dfrac{2\cos f_{2k}(y)\cos f_{2l}(y)}{(t^2+f^2_{2k}(y))(t^2+f^2_{2l}(y))}[(t^2-y^2)-e^{-2t}(t-y)^2], & k \neq l. \end{cases} \tag{13}$$

PROOF. Using some computations of Lemma 2, we obtain

$$\int_{-1}^{1}\int_{-1}^{1} e^{-t|x-x'|}U_{2k}^{y}(x)U_{2l}^{y}(x')\,dx'\,dx$$
$$= 2\left(\frac{t\int_{-1}^{1} U_{2k}^{y}(x)U_{2l}^{y}(x)\,dx}{t^2+f_{2l}^2(y)} - \frac{e^{-t}(t-y)\cos f_{2l}(y)}{2(t^2+f_{2l}^2(y))}\int_{-1}^{1}(e^{tx}-e^{-tx})\cos f_{2k}(y)x\,dx\right);$$

We remark that

$$\int_{-1}^{1} U_{2k}^{y}(x)U_{2l}^{y}(x)\,dx = \delta_{k,l}H_k^2(y).$$

In order to prove the lemma, we need to show that the second summand in the parenthesis is equal to

$$\frac{\cos f_{2l}(y)\cos f_{2k}(y)}{(t^2+f_{2k}^2(y))(t^2+f_{2l}^2(y))}[(t^2-y^2)-e^{-2t}(t-y)^2].$$

This is established as follows:

$$e^{-t}\int_{-1}^{1}(e^{tx}+e^{-tx})\cos(f_{2k}(y)x)\,dx$$
$$= e^{-t}\left[\frac{e^{tx}}{t^2+f_{2k}^2(y)}(t\cos(f_{2k}(y)x)+f_{2k}(y)\sin f_{2k}(y)x)|_{-1}^{1} + \frac{e^{-tx}}{t^2+f_{2k}^2(y)}(-t\cos(f_{2k}(y)x)+f_{2k}(y)\sin(f_{2k}(y)x))|_{-1}^{1}\right]$$
$$= \frac{2}{t^2+f_{2k}^2(y)}[(t\cos f_{2k}(y)+f_{2k}(y)\sin f_{2k}(y)) - e^{-2t}(t\cos f_{2k}(y)-f_{2k}(y)\sin f_{2k}(y))].$$

Since $y\cos f_{2k}(y) = f_{2k}(y)\sin f_{2k}(y)$, we obtain

$$-\frac{e^{-t}(t-y)\cos f_{2l}(y)}{2(t^2+f_{2l}^2(y))}\int_{-1}^{1}(e^{tx}+e^{-tx})\cos(f_{2k}(y)x)\,dx$$
$$= -\frac{\cos f_{2l}(y)\cos f_{2k}(y)}{(t^2+f_{2l}^2(y))(t^2+f_{2k}^2(y))}[(t^2-y^2)-e^{-2t}(t-y)^2].\quad \text{Q.E.D.}$$

COROLLARY 1.

$$\int_{-1}^{1}\int_{-1}^{1} e^{-t|x-x'|}U_{2k}^{0}(x)U_{2l}^{0}(x')\,dx'\,dx$$
$$= \int_{-1}^{1}\int_{-1}^{1} e^{-t|x-x'|}\cos(k\pi x)\cos(l\pi x')\,dx'\,dx$$
$$= \begin{cases} 2\left(\dfrac{t(\delta_{k,0}+1)}{t^2+(k\pi)^2} - \dfrac{t^2(1-e^{-2t})}{(t^2+(k\pi)^2)^2}\right), & \textit{for } k=l; \\[2ex] -\dfrac{2(-1)^k(-1)^l t^2(1-e^{-2t})}{(t^2+(k\pi)^2)(t^2+(l\pi)^2)}, & \textit{for } k\neq l. \end{cases}$$

We remark that $H_{2k}^2(0) = 1 + \delta_{k,0}$, where $\delta_{k,0}$ is the Kronecker symbol.

Corollary 1 can be obtained as a consequence of the previous lemma, or through arguments similar to those of Lemma 4.

COROLLARY 2.

$$\int_{-1}^{1}\int_{-1}^{1} e^{-t|x-x'|} U_{2k}^{\infty}(x) U_{2l}^{\infty}(x')\,dx'\,dx$$
$$= \int_{-1}^{1}\int_{-1}^{1} e^{-t|x-x'|}\cos(k+\tfrac{1}{2})\pi x\cos(l+\tfrac{1}{2})\pi x'\,dx'\,dx$$
$$= \begin{cases} 2\left(\dfrac{t}{t^2+(k+\frac{1}{2})^2\pi^2} + \dfrac{(k+\frac{1}{2})^2\pi^2(1+e^{-2t})}{(t^2+(k+\frac{1}{2})^2\pi^2)^2}\right) & \textit{for } k = l,\\ 2\dfrac{(1+e^{-2t})(k+\frac{1}{2})\pi^2(l+\frac{1}{2})(-1)^k(-1)^l}{(t^2+(k+\frac{1}{2})^2\pi^2)(t^2+(l+\frac{1}{2})^2\pi^2)}, & \textit{for } k \neq l. \end{cases}$$

Corollary 2 follows from (13), since

$$y\cos f_{2k}(y) = f_{2k}(y)\sin f_{2k}(y).$$

For each specific $U_{2k}^{\infty}(x)$ one can use arguments similar to those in the proof of the lemma to prove Corollary 2.

LEMMA 5.

$$\int_{-1}^{1}\int_{-1}^{1} e^{-t|x-x'|} U_{2k+1}^{y}(x) U_{2l+1}^{y}(x')\,dx'\,dx$$
$$= \begin{cases} 2\left(\dfrac{tH_{2k+1}^2(y)}{t^2+\psi_{2k+1}^2(y)} - \dfrac{\sin^2\psi_{2k+1}(y)}{(t^2+\psi_{2k+1}^2(y))^2}[(t^2-y^2)+e^{-2t}(t-y)^2]\right), & \\ & \textit{for } l = k,\\ -\dfrac{2\sin\psi_{2k+1}(y)\sin\psi_{2l+1}(y)}{(t^2+\psi_{2k+1}^2(y))(t^2+\psi_{2l+1}^2(y))}[(t^2-y^2)+e^{-2t}(t-y)^2], & \\ & \textit{for } l \neq k. \end{cases} \tag{14}$$

PROOF. Arguments similar to those in the proof of the previous lemma yield

$$\int_{-1}^{1}\int_{-1}^{1} e^{-t|x-x'|} U_{2k+1}^{y}(x) U_{2l+1}^{y}(x')\,dx'\,dx$$
$$= 2\left(\frac{t\int_{-1}^{1} U_{2k+1}^{y}(x) U_{2l+1}^{y}(x)\,dx}{t^2+\psi_{2l+1}^2(y)}\right.$$
$$\left. - \frac{e^{-t}(t-y)\sin\psi_{2l+1}(y)}{2(t^2+\psi_{2l+1}^2(y))}\int_{-1}^{1}(e^{tx}-e^{-tx})\sin(\psi_{2k+1}(y)x)\,dx\right)$$

and

$$\int_{-1}^{1}(e^{tx}-e^{-tx})\sin\psi_{2k+1}(y)x\,dx$$
$$=\frac{e^{tx}}{t^2+\psi_{2k+1}^2(y)}(t\sin\psi_{2k+1}(y)x-\psi_{2k+1}(y)\cos\psi_{2k+1}(y)x)|_{-1}^{1}$$
$$+\frac{e^{-tx}}{t^2+\psi_{2k+1}^2(y)}(t\sin\psi_{2k+1}(y)x+\psi_{2k+1}(y)\cos\psi_{2k+1}(y)x)|_{-1}^{1}$$
$$=\frac{2e^{t}}{t^2+\psi_{2k+1}^2(y)}[(t+y)+e^{-2t}(t-y)].$$

This proves Lemma 5.

COROLLARY 1.

$$\int_{-1}^{1}\int_{-1}^{1}e^{-t|x-x'|}U_{2k+1}^{0}(x)U_{2l+1}^{0}(x')\,dx'\,dx$$
$$=\int_{-1}^{1}\int_{-1}^{1}\sin((k+\tfrac{1}{2})\pi x)e^{-t|x-x'|}\sin((l+\tfrac{1}{2})\pi x')\,dx'\,dx$$
$$=\begin{cases}2\left(\dfrac{t}{t^2+(k+\frac{1}{2})^2\pi^2}-\dfrac{t^2(1+e^{-2t})}{(t^2+(k+\frac{1}{2})^2\pi^2)^2}\right), & \textit{for } k=l,\\[2ex] -\dfrac{2t^2(-1)^l(1+e^{-2t})(-1)^k}{(t^2+(k+\frac{1}{2})^2\pi^2)(t^2+(l+\frac{1}{2})^2\pi^2)}, & \textit{for } k\neq l.\end{cases}$$

COROLLARY 2.

$$\int_{-1}^{1}\int_{-1}^{1}e^{-t|x-x'|}U_{2k+1}^{\infty}(x)U_{2l+1}^{\infty}(x')\,dx'\,dx$$
$$=\int_{-1}^{1}\int_{-1}^{1}e^{-t|x-x'|}\sin((k+1)\pi x)\sin((l+1)\pi x')\,dx'\,dx$$
$$=\begin{cases}2\left(\dfrac{t}{t^2+(k+1)^2\pi^2}+\dfrac{(k+1)^2\pi^2(1-e^{-2t})}{(t^2+(k+1)^2\pi^2)^2}\right), & \textit{for } k=l;\\[2ex] 2\dfrac{(1-e^{-2t})(k+1)\pi(-1)^{k+1}(l+1)\pi(-1)^{l+1}}{(t^2+(k+1)^2\pi^2)(t^2+(l+1)^2\pi^2)}, & \textit{for } k\neq l.\end{cases}$$

Corollary 2 follows from (11), since

$$y\sin\psi_{2k+1}(y)=-\psi_{2k+1}(y)\cos\psi_{2k+1}(y).$$

4. In what follows we will use thc minimax variational definition of eigenvalues [6].

Given an operator L in a Hilbert space H, with eigenvalues indexed in decreasing order ($\eta_0\geq\eta_1\geq\cdots$), the kth eigenvalue is defined by

$$\eta_k=\inf_{v_1,\dots,v_k\in H}\sup(Lu,u)/(u,u),$$

where the supremum is over $\{u\in H\colon (u,v_i)=0,\ i=1,\dots,k\}$. In particular, if $v_1,\dots,v_k$ are chosen to be $u_0,\dots,u_{k-1}$ (the eigenfunctions corresponding to $\eta_0, \dots, \eta_{k-1}$), then we obtain a recursive variational definition of eigenvalues:

$$\eta_k = \sup(Lu, u)/(u, u),$$

where the supremum is over $\{u \in H\colon (u, u_j) = 0,\ j = 0, \dots, k-1\}$.

Before proceeding to the proof of our main theorem we introduce some definitions and notation. Since the eigenfunctions of equation (4) divide into two subspaces (of even functions and odd functions), and the system of eigenfunctions is complete in $L_2[-1,1]$, it follows that $L_2[-1,1]$ divides into two subspaces (the even ($L_2^{\text{e}}[-1,1]$) and the odd ($L_2^{\text{o}}[-1,1]$) functions in $L_2[-1,1]$):

$$L_2^{\text{e}}[-1,1] \oplus L_2^{\text{o}}[-1,1] = L_2[-1,1].$$

Lemma 1 implies that each eigenfunction of equation (4) can be referred either to $L_2^{\text{e}}[-1,1]$ or to $L_2^{\text{o}}[-1,1]$. Therefore, we choose only even eigenfunctions of (4), i.e., $U_k(x) \in L_2^{\text{e}}[-1,1]$, and index them in order of increase of their corresponding eigenvalues. The eigenvalue corresponding to the lth even eigenfunction we denote by $1/\mu_l$.

Similarly, we renumber the odd eigenfunctions of equation (4). The eigenvalue corresponding to the lth odd eigenfunction we denote by $1/\eta_l$. Both in $L_2^{\text{e}}[-1,1]$ and in $L_2^{\text{o}}[-1,1]$ we choose two different bases, namely:

$$\cos f_{2k}(0)x \quad \text{and} \quad \cos f_{2k}(\infty)x \qquad \text{in } L_2^{\text{e}}[-1,1];$$
$$\sin \psi_{2k+1}(0)x \quad \text{and} \quad \sin \psi_{2k+1}(\infty)x \qquad \text{in } L_2^{\text{o}}[-1,1].$$

LEMMA 6. *If $f(x) = \sum_{k=0}^{\infty} C_{2k}\cos(k\pi x)$, then*

$$(Bf, f) = \sum_{k=0}^{\infty} \frac{2C_{2k}^2 t(\delta_{k,0}+1)}{t^2 + (k\pi)^2} - 2t^2(1 - e^{-2t})\left(\sum_{k=0}^{\infty} \frac{(-1)^k C_{2k}}{t^2 + (k\pi)^2}\right)^2. \tag{15}$$

If $f(x) = \sum_{k=0}^{\infty} C_{2k}\cos((k+\frac{1}{2})\pi x)$, then

$$(Bf, f) = \sum_{k=0}^{\infty} \frac{2tC_{2k}^2}{t^2 + (k+\frac{1}{2})^2\pi^2} + (1 + e^{-2t})\left(\sum_{k=0}^{\infty} \frac{(-1)^k C_{2k}(k\pi + \frac{\pi}{2})}{t^2 + (k\pi + \frac{\pi}{2})^2}\right)^2. \tag{16}$$

If $g(x) = \sum_{k=0}^{\infty} C_{2k+1}\sin((k+\frac{1}{2})\pi x)$, then

$$(Bg, g) = \sum_{k=0}^{\infty} \frac{2tC_{2k+1}^2}{t^2 + (k+\frac{1}{2})^2\pi^2} - 2t(1 + e^{-2t})\left(\sum_{k=0}^{\infty} \frac{(-1)^k C_{2k+1}}{t^2 + (k+\frac{1}{2})^2\pi^2}\right)^2. \tag{17}$$

If $g(x) = \sum_{k=0}^{\infty} C_{2k+1}\sin((k+1)\pi x)$, then

$$(Bg, g) = \sum_{k=0}^{\infty} \frac{2tC_{2k+1}^2}{t^2 + (k+1)^2\pi^2} + 2(1 - e^{-2t})\left(\sum_{k=0}^{\infty} \frac{(-1)^{k+1} C_{2k+1}(k+1)\pi}{t^2 + (k+1)^2\pi^2}\right)^2. \tag{18}$$

Here

$$(Bf, f) = \int_{-1}^{1}\int_{-1}^{1} e^{-t|x-x'|} f(x) f(x')\, dx'\, dx.$$

PROOF. The proof will be carried out only for equality (15), since it is similar for the others.

Corollary 1 of Lemma 4 implies the chain of equalities

$$(Bf,f) = \int_{-1}^{1}\int_{-1}^{1} e^{-r|x-x'|}\left(\sum_{k=0}^{\infty} C_{2k}\cos(k\pi x)\right)\left(\sum_{l=0}^{\infty} C_{2l}\cos(l\pi x')\right) dx'\,dx$$
$$= \sum_{l=0}^{\infty}\sum_{k=0}^{\infty} C_{2k}C_{2l}\int_{-1}^{1}\int_{-1}^{1} e^{-t|x-x'|}\cos(k\pi x)\cos(l\pi x')\,dx'\,dx$$
$$= -2\sum_{l=0}^{\infty}\sum_{k=0}^{\infty} C_{2k}C_{2l}\frac{(-1)^k(-1)^l}{(t^2+(k\pi)^2)(t^2+(l\pi)^2)}t^2(1-e^{-2t}) + 2\sum_{k=0}^{\infty}\frac{t(\delta_{k0}+1)}{t^2+(k\pi)^2}$$
$$= 2\left(\sum_{k=0}^{\infty}\frac{t(\delta_{k,0}+1)}{t^2+(k\pi)^2} - t^2(1-e^{-2t})\left(\sum_{k=0}^{\infty}\frac{(-1)^k C_{2k}}{t^2+(k\pi)^2}\right)^2\right).$$

Before proceeding to the proof of the next theorems, we introduce notation which will simplify writing of some expressions.

Let $v_0,\dots,v_k$ be arbitrary elements of some Hilbert space H. Then the subspace orthogonal to these elements will be denoted by

$$H(v_0,v_1,\dots,v_k) = \{u\in H | (u,v_0)_H = 0,\dots,(u,v_k)_H = 0\}.$$

THEOREM 2. *All eigenvalues* μ_l *and* η_l *satisfy* $\mu_l > \eta_l > \mu_{l+1}$, $l = 0,1,\dots$.

PROOF. We have

$$\mu_l = \inf_{v_0,\dots,v_{l-1}\in L_2^e}\ \sup_{\substack{u\in L_2^e(v_0,\dots,v_{l-1})\\ \|u\|=1}} \int_1^{\infty}\frac{1}{t}\int_{-1}^{1}\int_{-1}^{1} e^{-t|x-x'|}u(x)u(x')\,dx'\,dx\,dt$$
$$= \sup_{\substack{u\in L_2^e(u_0^e,\dots,u_{l-1}^e)\\ \|u\|=1}} \int_1^{\infty}\frac{1}{t}\int_{-1}^{1}\int_{-1}^{1} e^{-t|x-x'|}u(x)u(x')\,dx'\,dx\,dt. \tag{19}$$

Here u_i^e is the ith even eigenfunction. We expand $u_i^e(x)$, $i=0,\dots,l$, in the basis $\cos((k+\frac{1}{2})\pi x)$, $k=0,1,\dots$,

$$u_i^e(x) = \sum_{k=0}^{\infty} C_{2k}^i\cos((k+\tfrac{1}{2})\pi x)$$

and, after this substitution, (18) yields

$$\mu_1 = \sup_{\substack{c\in l_2(c^0,\dots,c^{l-1})\\ \|c\|_{l_2}=1}} 2\left(\int_1^\infty \sum_{k=0}^\infty \frac{c_{2k}^2}{t^2+(k+\frac{1}{2})^2\pi^2}\,dt\right.$$

$$\left. + \int_1^\infty \frac{1+e^{-2t}}{t}\left(\sum_{k=0}^\infty \frac{c_{2k}(-1)^k(k+\frac{1}{2})\pi}{t^2+(k\pi+\frac{\pi}{2})^2}\right)^2 dt\right)$$

$$> \sup_{\substack{c\in l_2(c^0,\dots,c^{l-1})\\ \|c\|_{l_2}=1}} 2\left(\int_1^\infty \sum_{k=0}^\infty \frac{c_{2k}^2}{t^2+(k+\frac{1}{2})^2\pi^2}\,dt\right.$$

$$\left. - \int_1^\infty t(1+e^{-2t})\left(\sum_{k=0}^\infty \frac{c_{2k}(-1)^k}{t^2+(k+\frac{1}{2})^2\pi^2}\right)^2 dt\right).$$

Here l_2 is the Hilbert space of sequences, and

$$c^i = (c_0^i, c_2^i, c_4^i, \dots), \qquad c = (c_0, c_2, c_4, \dots).$$

We denote

$$v_i^{\circ}(x) = \sum_{k=0}^\infty c_{2k}^i \sin(k+\tfrac{1}{2})\pi x.$$

Thus, applying (17), we obtain

$$\mu_l > \sup_{\substack{u\in L_2^{\circ}(v_0^{\circ},\dots,v_{l-1}^{\circ})\\ \|u\|=1}} \int_1^\infty \frac{1}{t}\int_{-1}^1\int_{-1}^1 e^{-t|x-x'|}u(x)u(x')\,dx'\,dx\,dt$$

$$\geq \inf_{v_0,\dots,v_{l-1}\in L_2^{\circ}[-1,1]} \sup_{\substack{u\in L_2^{\circ}(v_0,\dots,v_{l-1})\\ \|u\|=1}} \int_{-1}^1 \frac{1}{t}\int_{-1}^1\int_{-1}^1 e^{-t|x-x'|}u(x)u(x')\,dx'\,dx\,dt$$

$$= \eta_l.$$

This proves the first part of the inequality. Let us prove the second:

$$\eta_1 = \inf_{v_0,\dots,v_{l-1}\in L_2^{\circ}[-1,1]} \sup_{\substack{u\in L_2^{\circ}(v_0,\dots,v_{l-1})\\ \|u\|=1.}} \int_1^\infty \frac{1}{t}\int_{-1}^1\int_{-1}^1 e^{-t|x-x'|}u(x)u(x')\,dx'\,dx\,dt$$

$$= \sup_{\substack{u\in L_2^{\circ}(u_0^{\circ},\dots,u_{l-1}^{\circ})\\ \|u\|=1}} \int_1^\infty \frac{1}{t}\int_{-1}^1\int_{-1}^1 e^{-t|x-x'|}u(x')u(x)\,dx'\,dx\,dt. \tag{20}$$

Here $u_i^{\circ}(x)$ is the ith odd eigenfunction. We expand the $u_i(x)$, $i=0,\dots,l-1$, in the basis $\sin((k+1)\pi x)$, $k=0,1,\dots$, of $L_2^{\circ}[-1,1]$. Then, substituting,

$$u_i^{\circ}(x) = \sum_{k=0}^\infty c_{k+1}^i \sin((k+1)\pi x)$$

into (20), we get

$$\eta_l = \sup_{\substack{c\in l_2(c^0,\dots,c^{l-1})\\ \|c\|_{l_2}=1}} 2\left(\int_1^\infty \sum_{k=0}^\infty \frac{c_{k+1}^2}{t^2+(k+1)^2\pi^2}\,dt\right.$$

$$\left. + \int_1^\infty \frac{1-e^{-2t}}{t}\left(\sum_{k=0}^\infty \frac{(k+1)\pi c_{k+1}(-1)^{k+1}}{t^2+(k+1)^2\pi^2}\right)^2 dt\right)$$

$$> \sup_{\substack{c\in l_2(c^0,\dots,c^{l-1})\\ \|c\|_{l_2}=1}} 2\left(\int_1^\infty \sum_{k=1}^\infty \frac{c_k^2}{t^2+k^2\pi^2}\,dt\right.$$

$$\left. - \int_1^\infty t(1-e^{-2t})\left(\sum_{k=1}^\infty \frac{c_k(-1)^k}{t^2+(k\pi)^2}\right)^2 dt\right) = A.$$

Here $c = (c_1, c_2, \dots)$ and $c^i = (c_1^i, c_2^i, \dots)$ are elements of l_2.

We denote

$$v_{i+1}^{\mathrm{e}}(x) = \sum_{k=1}^\infty c_k^i \cos k\pi x, \qquad i = 0, \dots, l-1,$$

$$u(x) = \sum_{k=1}^\infty c_k \cos k\pi x.$$

This shows that all these functions are orthogonal to the function $\cos 0 = 1$. Thus, (15) yields

$$\eta_l > A = \sup_{\substack{u\in L^{\mathrm{e}}(1,v_1^{\mathrm{e}},\dots,v_l^{\mathrm{e}})\\ \|u\|=1}} \int_1^\infty \frac{1}{t}\int_{-1}^1\int_{-1}^1 e^{-t|x-x'|}u(x)u(x')\,dx'\,dx\,dt$$

$$\geq \inf_{v_0,\dots,v_l\in L_2^{\mathrm{e}}[-1,1]} \sup_{\substack{u\in L_2^{\mathrm{e}}(v_0,\dots,v_l)\\ \|u\|=1}} \int_1^\infty \frac{1}{t}\int_{-1}^1\int_{-1}^1 e^{-t|x-x'|}u(x)u(x')\,dx'\,dx\,dt = \mu_{l+1}.$$

Q.E.D.

COROLLARY. *All the eigenvalues of equation* (4) *are simple, and* λ_{2k} *corresponds to an even eigenfunction, while* λ_{2k+1} *corresponds to an odd.*

THEOREM 3. *For all* $k = 1, 2, \dots$,

$$\int_1^\infty \frac{2}{t^2+(k\pi/2)^2}\,dt > \frac{1}{\lambda_k} > \int_1^\infty \frac{2}{t^2+((k+1)\pi/2)^2}\,dt. \tag{21}$$

REMARK.

$$\int_1^\infty \frac{1}{t^2+(k\pi/2)^2}\,dt = \begin{cases} 1, & k=0,\\ \frac{2}{k\pi}\left(\frac{\pi}{2} - \operatorname{Arctan}\frac{2}{k\pi}\right), & k>0. \end{cases}$$

PROOF. We use arguments similar to those of the previous theorem, but only for the purpose of obtaining (21). Let k be an even number (for k odd the proof is similar), $k = 2l$. We use (19), and, as in Theorem 2, we expand all the functions in the basis $\cos((k+\frac{1}{2})\pi x)$, $k = 0, 1, \dots$, of $L_2^{\mathrm{e}}[-1,1]$. This results in the following chain of inequalities:

$$\begin{aligned}
\frac{1}{\lambda_k} = \mu_l = &\sup_{\substack{c\in l_2(c^0,\dots,c^{l-1})\\ \|c\|_{l_2}=1}} 2\left(\int_1^\infty \sum_{k=0}^\infty \frac{c_{2k}^2}{t^2+(k+\frac{1}{2})^2\pi^2}\,dt\right.\\
&\left.+\int_1^\infty \frac{1+e^{-2t}}{t}\left(\sum_{k=0}^\infty \frac{c_{2k}(-1)^k(k+\frac{1}{2})\pi}{t^2+(k\pi+\frac{\pi}{2})^2}\right)^2 dt\right)\\
> &\sup_{\substack{c\in l_2(c^0,\dots,c^{l-1})\\ \|c\|_{l_2}=1}} 2\int_1^\infty \sum_{k=0}^\infty \frac{c_{2k}^2}{t^2+(k+\frac{1}{2})^2\pi^2}\,dt\\
\geq &\inf_{v_0,\dots,v_{l-1}\in L_2^{\mathrm{e}}[-1,1]} \sup_{\substack{u\in L_2^{\mathrm{e}}(v_0,\dots,v_{l-1})\\ \|u\|=1}} \int_1^\infty \sum_{k=0}^\infty \frac{c_{2k}^2}{t^2+(k+\frac{1}{2})^2\pi^2}\,dt\\
= &\int_1^\infty \frac{2}{t^2+(l+\frac{1}{2})^2\pi^2}\,dt = \int_1^\infty \frac{2}{t^2+((k+1)\pi/2)^2}\,dt.
\end{aligned}$$

To prove the left-hand side of (21), we expand all the functions of (19) in the basis $\cos(k\pi x)$, $k = 0, 1, \dots$, of $L_2^{\mathrm{e}}[-1,1]$. Then for all $v_i \in L_2^{\mathrm{e}}[-1,1]$ we have

$$v_i^{\mathrm{e}}(x) = \sum_{k=0}^\infty v_i^{\mathrm{e}}(2k)\cos(k\pi x). \tag{22}$$

The substitution of (22) into the first part of (19) yields

$$\begin{aligned}
\mu_l = &\inf_{c^0,\dots,c^{l-1}\in l_2} \sup_{\substack{c\in l_2(c^0,\dots,c^{l-1})\\ \|c\|_{l_2}=1}} 2\left(\int_1^\infty \sum_{k=0}^\infty \frac{c_{2k}^2}{t^2+(k\pi)^2}\,dt\right.\\
&\left.-\int_1^\infty t(1-e^{-2t})\left(\sum_{k=0}^\infty \frac{c_{2k}(-1)^k}{t^2+(k\pi)^2}\right)^2 dt\right)\\
< &\inf_{c^0,\dots,c^{l-1}\in l_2} \sup_{\substack{c\in l_2(c^0,\dots,c^{l-1})\\ \|c\|_{l_2}=1}} 2\int_1^\infty \sum_{k=0}^\infty \frac{c_{2k}^2}{t^2+(k\pi)^2}\,dt\\
= &\int_1^\infty \frac{2}{t^2+(l\pi)^2}\,dt = \int_1^\infty \frac{2}{t^2+(k\pi/2)^2}\,dt.
\end{aligned}$$

Thus, for k even inequality (21) is proved. For k odd it is proved similarly by decomposing with respect to two different bases: $\sin((k+\frac{1}{2})\pi x)$ and $\sin((k+1)\pi x)$ of $L_2^{\mathrm{o}}(-1,1]$.

Thus Theorem 3 is proved.

For λ_0 and λ_1 these two-sided estimates can be sharpened by decreasing the upper bound.

THEOREM 4.

$$\int_1^\infty \frac{2}{t+f_0^2(t)}\,dt > \frac{1}{\lambda_0}, \tag{23}$$

$$\int_1^\infty \frac{2}{t^2+\psi_1^2(t)}\,dt > \frac{1}{\lambda_1}. \tag{24}$$

PROOF. By the corollary of Theorem 2, λ_0 corresponds to an even eigenfunction. By definition,

$$\begin{aligned}
\frac{1}{\lambda_0} &= \sup_u \int_1^\infty \frac{1}{t} \int_{-1}^1 \int_{-1}^1 e^{-t|x-x'|} u(x)u(x')\,dx'\,dx\,dt \\
&= \int_1^\infty \frac{1}{t} \int_{-1}^1 \int_{-1}^1 e^{-t|x-x'|} u_0(x)u_0(x')\,dx'\,dx\,dt \\
&\le \int_1^\infty \frac{1}{t} \sup_{u_t(x)} \int_{-1}^1 \int_{-1}^1 e^{-t|x-x'|} u_t(x)u_t(x')\,dx'\,dx\,dt \\
&= \int_1^\infty \frac{\mu_0(t)}{t}\,dt = \int_1^\infty \frac{2}{t^2+f_0^2(t)}\,dt.
\end{aligned}$$

In the case of equality, for all t we would have

$$\begin{aligned}
&\int_{-1}^1 \int_{-1}^1 e^{-t|x-x'|} u_0(x)u_0(x')\,dx'\,dx \\
&\quad = \sup_{u_t(x)} \int_{-1}^1 \int_{-1}^1 e^{-t|x-x'|} u_t(x)u_t(x')\,dx'\,dx \\
&\quad = \int_{-1}^1 \int_{-1}^1 e^{-t|x-x'|} \cos(f_0(t)x)\cos(f_0(t)x')\,dx'\,dx.
\end{aligned}$$

This implies that $u_0(x)$ is an eigenfunction of the operator L and $u_0(x) \equiv \cos f_0(t)x$; but $u_0(x)$ does not depend on t. Therefore, the inequality is strict.

We will prove the second inequality.

The corollary to Theorem 2 implies that the eigenvalue λ_1 corresponds to an odd eigenfunction:

$$\begin{aligned}
\frac{1}{\lambda_1} &= \sup_{u\in L_2^{\circ}[-1,1]} \int_1^\infty \frac{1}{t} \int_{-1}^1 \int_{-1}^1 e^{-t|x-x'|} u(x)u(x')\,dx'\,dx\,dt \\
&= \int_1^\infty \frac{1}{t} \int_{-1}^1 \int_{-1}^1 e^{-t|x-x'|} u_1(x)u_1(x')\,dx'\,dx\,dt \\
&\le \int_1^\infty \frac{1}{t} \sup_{\substack{u\in L_2(\cos f_0(t)x)\\ \|u\|=1}} \int_{-1}^1 \int_{-1}^1 e^{-t|x-x'|} u(x)u(x')\,dx'\,dx\,dt \\
&= \int_1^\infty \frac{1}{t} \sup_{\substack{u\in L_2^{\circ}[-1,1]\\ \|u\|=1}} \int_{-1}^1 \int_{-1}^1 e^{-t|x-x'|} u(x)u(x')\,dx'\,dx\,dt \\
&= \int_1^\infty \frac{\mu_1(t)}{t}\,dt = \int_1^\infty \frac{2}{t^2+\psi_1^2(t)}\,dt.
\end{aligned}$$

Since $u_1(x)$ does not depend on t, we see that equality does not take place, and therefore the inequality is proved. One could argue that (23) and (24) are not estimates, since $f_0(t)$ and $\psi_1(t)$ are in fact unknown functions. But it turns out that the integrals

$$A = \int_1^\infty \frac{2}{t^2 + f_0^2(t)}\,dt \quad \text{and} \quad B = \int_1^\infty \frac{2}{t^2 + \psi_1^2(t)}\,dt$$

can be expressed in terms of elementary functions and cosine integrals.

To show this, let $\alpha \tan \alpha = 1$ and $\alpha \in [0, \pi/2)$, while $-\beta \cot \beta = 1$ and $\beta \in [\pi/2, \pi)$. In the first integral we make the change of variable $\tau = f_0(t)$, $\tau \in [\alpha, \pi/2]$. Then

$$\begin{aligned} A &= 2\int_\alpha^{\pi/2} \frac{d(\tau \tan \tau)}{\tau^2 (1/\cos^2 \tau)} = 2\left[\int_\alpha^{\pi/2} \frac{d\tau}{\tau} + \int_\alpha^{\pi/2} \frac{\sin \tau \cos \tau}{\tau^2}\,d\tau\right] \\ &= 2(\ln \tfrac{\pi}{2} - \ln \alpha) + 2\int_{2\alpha}^{\pi} \frac{\sin x}{x^2}\,dx \\ &= 2(\ln \tfrac{\pi}{2} - \ln \alpha) + \frac{\sin 2\alpha}{\alpha} + 2(\mathrm{ci}(\pi) - \mathrm{ci}(2\alpha)). \end{aligned}$$

Here $\mathrm{ci}(x) = \int x^{-1} \cos x\,dx$ (cosine integral). The change of variable $\tau = \psi_1(t)$, $\tau \in [\beta, \pi]$, in the second integral gives us

$$\begin{aligned} B &= 2\int_\beta^{\pi} \frac{d(-\tau \cot \tau)}{\frac{\tau^2}{\sin^2 \tau}} = 2(\ln \pi - \ln \beta) - \int_\beta^{\pi} \frac{\sin 2\tau}{\tau^2}\,d\tau \\ &= 2(\ln \pi - \ln \beta) - \frac{\sin 2\beta}{\beta} - 2(\mathrm{ci}(2\pi) - \mathrm{ci}(2\beta)). \end{aligned}$$

5. We conclude by mentioning that all the results are obtained for equation (1) with the boundary conditions (2), which is a particular case of the equation

$$\mu \frac{\partial \varphi}{\partial x} + \alpha \varphi(x, \mu) = \frac{\lambda}{2} \int_{-1}^{1} \varphi(x, \mu)\,d\mu$$
$$(x \in [-a, a],\ \mu \in [-1, 1],\ a < \infty,\ 0 < \alpha = \text{const} < \infty) \tag{25}$$

with boundary conditions

$$\varphi(a, \mu) = 0, \quad \mu < 0, \qquad \varphi(-a, \mu) = 0, \quad \mu > 0. \tag{26}$$

However, all our results are also valid for problem (25), (26). In this connection Theorem 1 is restated as follows.

THEOREM 1′. *The integral equation*

$$\mu_k(t) u_k^t(x) = \int_{-a}^{a} e^{-\alpha t |x - x'|} u_k^t(x')\,dx', \tag{27}$$

where $0 < \alpha < \infty$, *has eigenfunctions*

$$u_{2k}^t(x) = \cos f_{2k}(\alpha t a)\frac{x}{a}, \quad k = 0, 1, \dots, \tag{28}$$

$$u_{2k+1}^t(x) = \sin \psi_{2k+1}(\alpha t a)\frac{x}{a}, \quad k = 0, 1, \dots, \tag{29}$$

and their corresponding eigenvalues are

$$\mu_{2k}(t) = \frac{2ta^2\alpha}{t^2a^2\alpha^2 + f_{2k}^2(ta\alpha)}, \tag{30}$$

$$\mu_{2k+1}(t) = \frac{2ta^2\alpha}{t^2a^2\alpha^2 + \psi_{2k+1}^2(ta\alpha)}, \tag{31}$$

where f_{2k} *and* ψ_{2k+1} *are defined in Lemmas* 2 *and* 3. *Equation* (27) *has no other eigenvalues.*

Theorem 2 and its corollary remain unchanged.

Theorems 3 and 4 in this case are restated as follows.

THEOREM 3′. *For all* $k = 0, 1, 2, \ldots,$

$$2a\int_{a\alpha}^{\infty}\frac{dt}{t^2 + (k\pi/2)^2} > \frac{1}{\lambda_k} > 2a\int_{a\alpha}^{\infty}\frac{dt}{t^2 + ((k+1)\pi/2)^2}. \tag{32}$$

REMARK.

$$\int_{a\alpha}^{\infty}\frac{dt}{t^2 + (k\pi/2)^2} = \begin{cases} \dfrac{1}{a\alpha} & \text{for } k = 0, \\ \dfrac{2}{k\pi}\left(\dfrac{\pi}{2} - \operatorname{Arctan}\dfrac{2a\alpha}{k\pi}\right) & \text{for } k > 0. \end{cases}$$

THEOREM 4′. *The eigenvalues of problem* (25), (26) *satisfy*

$$2a\int_{a\alpha}^{\infty}\frac{1}{t^2 + f_0^2(t)}\,dt > \frac{1}{\lambda_0}, \tag{33}$$

$$2a\int_{\alpha a}^{\infty}\frac{1}{t^2 + \psi_1^2(t)}\,dt > \frac{1}{\lambda_1}. \tag{34}$$

6. A more general theorem can be formulated; its proof repeats all the arguments of this paper.

THEOREM 5. *Consider the equation*

$$\Phi_k(x) = \lambda_k\int_{-a}^{a}\left(\int_1^{\infty}e^{-t\alpha|x-x'|}m(t)\,dt\right)\Phi_k(x')\,dx'$$

$$(0 < a < \infty,\ 0 < \alpha = \text{const} < \infty), \tag{35}$$

where $m(t) \geq 0$ *is a function defined in* $(0, \infty)$ *satisfying the following conditions:*

A. $\int_1^{\infty}e^{-t\alpha|x-x'|}m(t)\,dt \in L_2([-a,a]\times[-a,a])$.

B. *For all* $\varphi(x') \in L_2[-a,a]$ *and almost all* $x \in [-a,a]$

$$\int_{-a}^{a}\left(\int_1^{\infty}e^{-\alpha t|x-x'|}m(t)\,dt\right)\varphi(x')\,dx' = \int_1^{\infty}\left(\int_{-a}^{a}e^{-\alpha t|x-x'|}\varphi(x')\,dx'\right)m(t)\,dt.$$

Then the following assertions are true:

1) *The eigenvalues of problem* (35) *are simple.*

2) *The eigenvalues of problem* (35) *satisfy*

$$2a\int_{\alpha a}^{\infty}\frac{tm(t)}{t^2+f_0^2(t)}\,dt>\frac{1}{\lambda_0}>2a\int_{\alpha a}^{\infty}\frac{tm(t)}{t^2+(\pi/2)^2}\,dt$$

$$>2a\int_{\alpha a}^{\infty}\frac{tm(t)}{t^2+\psi_1^2(t)}\,dt>\frac{1}{\lambda_1}>2a\int_{\alpha a}^{\infty}\frac{tm(t)}{t^2+\pi^2}\,dt$$

and

$$2a\int_{\alpha a}^{\infty}\frac{tm(t)}{t^2+(k\pi/2)^2}\,dt>\frac{1}{\lambda_k}>2a\int_{\alpha a}^{\infty}\frac{tm(t)}{t^2+((k+1)\pi/2)^2}\,dt,\quad k=2,3,\dots.$$

Here $f_0(t)$ *and* $\psi_1(t)$ *are the functions defined in Lemmas* 2 *and* 3.

The author is grateful to V. I. Agoshkov for checking over this paper and for a number of valuable remarks which helped to correct some inaccuracies. He also helped with the correct statement and the proof of Lemma 1.

The author is also grateful to A. G. Kleshchev for a discussion of this work.

BIBLIOGRAPHY

1. V. S. Vladimirov, *Mathematical problems of one-speed transport theory*, Trudy Mat. Inst. Steklov. **61** (1961). (Russian)

2. Herbert Bristol Dwight, *Tables of integrals and other mathematical data*, 4th ed., Macmillan, 1961.

3. F. R. Gantmakher and M. G. Kreĭn, *Oscillation matrices and kernels, and small oscillations of mechanical systems*, 2nd ed., GITTL, Moscow, 1950; German transl., Akademie-Verlag, Berlin, 1960; English transl., U.S. Atomic Energy Commission, Oak Ridge, Tenn., 1961.

4. G. I. Marchuk and V. I. Lebedev, *Numerical methods in neutron transport theory*, 2nd ed., Atomizdat, Moscow, 1981; English transl., Harwood Academic Publ., Chur, 1985.

5. G. Milton Wing, *An introduction to transport theory*, Wiley, 1962.

6. S. H. Gould, *Variational methods for eigenvalue problems*, 2nd rev. ed., Univ. of Toronto Press, Toronto, and Oxford Univ. Press, London, 1966.

Translated by V. Y. RICE

ABCDEFGHIJ 89